T0203350

RFID and the Internet of Things

RFID
and
the Internet of Things

Edited by
Hervé Chabanne
Pascal Urien
Jean-Ferdinand Susini

First published 2011 in Great Britain and the United States by ISTE Ltd and John Wiley & Sons, Inc.
Adapted and updated from *RFID et l'internet des choses* published 2010 in France by Hermes Science/Lavoisier © LAVOISIER 2010

ISTE Ltd
27-37 St George's Road
London SW19 4EU
UK

www.iste.co.uk

John Wiley & Sons, Inc.
111 River Street
Hoboken, NJ 07030
USA

www.wiley.com

© ISTE Ltd 2011

Library of Congress Cataloging-in-Publication Data

RFID et l'internet des choses. English
  RFID and the internet of things / edited by Herve Chabanne, Pascal Urien, Jean-Ferdinand Susini.
    p. cm.
  Includes bibliographical references and index.
  ISBN 978-1-84821-298-5
  1. Radio frequency identification systems. 2. Embedded Internet devices. I. Chabanne, Herve. II. Urien, Pascal. III. Susini, Jean-Ferdinand. IV. Title.
  TK6570.I34R479 2011
  384.6--dc22

                                                                              2011008134

British Library Cataloguing-in-Publication Data
A CIP record for this book is available from the British Library
ISBN 978-1-84821-298-5

Printed and bound in Great Britain by CPI Antony Rowe, Chippenham and Eastbourne.

MIX
Paper from
responsible sources
FSC
www.fsc.org
FSC® C013604

Table of Contents

Foreword

The RFID (*Radio Frequency Identification*) technology allows automatic identification of information contained in a tag by using radio waves. An RFID tag contains an antenna and a microchip to transmit and receive.

It appears as an alternative to barcodes that are facing the growth of the trade and new trade modes based on it. Indeed, if barcodes have proved, over a long period, their efficiency in data coding, they currently face some limitations such as the use of an optical reader (scanner) which has to be located at a relatively short distance from the identified object, or in a small data storage system.

The RFID technology is characterized by the deployment of three essential components: a microchip, an antenna and a reader. The tag is placed on the object or the person to be identified. It contains information that is decrypted by the server by using an antenna for transmitting signals between the reader and the chip. The radio frequencies used by the RFID technology are in the 50 kHz to 2.5 GHz range.

Therefore it is necessary to establish a comparison between the barcodes and the RFIDs, to understand why the RFIDs can replace the barcodes, and how they still have some limitations. The first difference between the two systems is their reading mode: a barcode is read by an optical laser, while an RFID tag is scanned by a reader that identifies the data contained in this tag. The reading distance of the RFID tags can be higher. Indeed, it extends from a few centimeters to 200 meters. In addition, the RFID tags can store more information than the barcodes, and the collected data can be up to several kilobytes. It is also important to note that the RFID tags can be recycled because new information can be registered in it. One of the drawbacks of the deployment of the RFID technology is its cost, which varies and can slow down its implementation, or perturbations between the tags and their sensitivity to interference waves.

The RFID technology is still a topic of great interest to many: it not only makes it possible to solve the problems faced by barcodes, but is also of importance to key sectors of economy and trade, such as distribution and transport. The implementation of this technology could revolutionize the pharmaceutical industry and represent a significant advance in the field of health which would benefit everyone: in fact, who would not expect to benefit from quality care and not like to avoid being a victim of medical errors? Applied to pharmaceutical products, smart tags would guarantee the authenticity and thus avoid counterfeiting. With regard to blood donations, their use would significantly reduce the risk of possible confusions. Medical staff would be able to authenticate the source of a blood sample without error, and thus lead to a correct transfusion.

It is clear that the RFID tags are more and more inescapable and deserve our attention. This was the main reason for us to write this book, which provides a description of the famous "smart tags" from a scientific viewpoint. It is divided into five parts: part 1 looks at the operation of the RFID systems. It establishes the classification of RFIDs, studies the physical aspect of tags and antennae, as well as coding techniques of RFID information. Part 2 is devoted to the application of RFID. It traces its evolution from the barcodes to the RFID tags by making a comparison between the two systems, and shows in a concrete manner various application examples of the RFID technology. Part 3 describes the cryptographic protocols of RFID.

The data contained in the tags must be identified without jeopardizing the privacy of the persons who possess them. Part 4 focuses on the global standardization of RFID: EPC (Electronic Product Code). It is a global architecture initialized by the Internet of objects and the desire to establish a large quantum of data for all products, while ensuring the specificity and authenticity of each one of them. And finally, part 5 attempts to describe the architecture to implement "the Internet of things" as efficiently as possible and by adapting to the evolution of needs: middlewares.

Guy PUJOLLE
April 2011

PART ONE

Physics of RFID

Chapter 1

Introduction

RFID (*Radio Frequency Identification*) systems use electromagnetic waves to transmit, at distance, energy and data to devices that perform a scheduled process of information contained in these exchanges. The origin of RFID technologies, dates back to the invention of RADAR, where, during the Second World War, fighter pilots cleverly maneuvered their planes to be remotely identified by friendly radar operators, who distinguished them from their foes (*Identify Friend or Foe*).

However, the RFID technology received a boost in the early 1970s. The very first RFID devices were simple resonant analog circuits. Then, advances in microelectronics allowed the integration of increasingly complex digital functions. The initial applications were designed to track and monitor dangerous materials in sensitive areas (usually military or nuclear). In the late 1970s, applications of these devices also included the civilian doman, typically the monitoring of animals, vehicles and automated production lines [DOB 07].

Usual tracking technologies such as barcodes, invented in 1970 by an IBM engineer, have shown their limitations for applications in an altered environment, such as animal tracking or in the engine assembly lines. Indeed, the barcode must pass through a scanning window to be scanned by a mobile reader without obstacles or of dirt traces which degrade or block the reading operation. This is why RFID technologies are being developed to replace barcodes in identification functions, so as to make it possible to read or write information at a distance using electromagnetic waves.

Chapter written by Simon ELRHARBI and Stefan BARBU.

RFID technologies and contactless smart cards consist of one or more electronic tags connected to one or more antennae or terminals that radiate an electromagnetic field through their antennae. These devices communicate by RF (Radio Frequency) or UHF (Ultra High Frequency) channels. Some RFID applications and contactless smart cards require embedded energy sources to facilitate the data exchange between the tags and the terminal readers. Other more common RFID technologies and contactless smart cards perform remote energy transmission to enable data exchange.

Depending on the frequencies used, the transmitted energy necessary for operations can be stored in a geometric volume, site of an electromagnetic induction effect (as in the case of low frequencies or radio frequencies), or propagated (as in the case of Ultra High Frequency).

The deployment of RFID technologies presupposes that a number of electrical electronic, mechanical and material parameters have been controlled [ELR 04].

Indeed, given the nature of energy and data transmissions between the devices of the RFID system, the geometric space in which the energy transmission and data exchanges are performed may reveal communication failures between tags and terminals. In particular, the phenomena of echo due to reflection and absorption signals (as in the case of UHF) must be controlled. The coupling intensity in the near field (as in case of LF or HF frequencies) can degrade the signal to noise ratio or, in other cases, lead to very high impedance mismatch at the power stage level and cause a malfunction of the baseband station [BAR 05]. The antenna orientation of RFID tag/terminal pairs in correlation with the energy and data propagation should be minimized and the writing and reading distance should be optimized by taking into account the complete front-end architecture (stages of power, reception and power control). These issues (operational zones and antenna configurations) mainly relate to physical and electrical properties.

Regarding electronic aspects, the anti-collision processing should allow us to establish data exchange between all the RFID tags in the operating zone of the terminal(s) under the conditions predefined by RFID system specifications and ISO standards. The processing time of transmission, which should be secured by cryptographic algorithms, should be optimized (in terms of bandwidth and data rates), so that the processing time is compatible with the high flux of expected transactions.

Finally, the interoperability between different RFID systems, their robustness (in terms of electrical features) and their compatibility with ISO standards must be guaranteed. The mechanical parameters (connections between chips and antennas) and materials (from a point of view of electric behaviors) are all very important in this context.

By nature of their capture and data processing abilities, the RFID technology is well suited for automation of the complete supply chain, with better utilization flexibility and operation under varying environmental constraints, even when the object is in movement and occupies various positions. RFID technologies follow the unprecedented development of international trade exchanges. These technologies make it possible to save money by avoiding logistical and human errors and by limiting fraud, irrespective of their origin [ROU 05].

Information system architectures that aggregate data from RFID systems are based on normative networks, which define international ISO standards, or the coalition of managers and assignees such as EPCglobal Inc. which implements EPC (Electronic Product Code) codes. These codes are allocated to objects for identification at a worldwide level, while providing an interconnection service to servers dedicated to identification and localization of objects by the Internet.

Today, due to the joint progress of micro-electronics, microcomputer and telecommunications, RFID systems are not only reserved for automatic identification, but have also spread to other areas such as secure access to buildings, to networks or the completion of secured transactions between remote electronic devices.

1.1. Bibliography

[BAR 05] BARBU S., Design and implementation of an RF metrology system for contactless identification systems at 13.56 MHz, PhD Thesis, University of Marne La Vallée, 2005.

[DOB 07] DOBKIN D.M., *The RF in RFID*, Elsevier, Oxford, 2007.

[ELR 04] ELRHARBI S., BARBU S., GASTON L., Why Class 1 PICC is not suitable for ICAO / NTWG E-passport - New proposal for a Class 1 PCD, Contribution ISO/IEC JTC1/SC17/WG8/TF2 num. N430, ISO/IEC JTC1/SC17/WG8/TF2, June 2004.

[ROU 05] ROURE F., GORICHON J., SARTORIUS E., RFID technologies: industrial and society issues, Report of CGTI committee num. Report N° II-B.9 - 2004, CGTI, January 2005.

Chapter 2

Characteristics of RFID Radio Signals

This chapter describes the main characteristics of electrical signals exchanged in RFID systems.

2.1. Description and operating principle of RFID systems

RFID systems typically consist of fixed elements (called basc station, reader, coupling device, terminal, etc.), whose function is to identify and process, by using radio waves, the information contained in one or more deported elements such as transponders, tags, badges, electronic tokens, or contactless smart cards. The different designations of the fixed and deported elements may vary, depending on applications, contexts, and their hardware and software resources. The fixed elements are themselves connected to servers for data processing at middleware and application levels for analysis, archiving and traceability (Figure 2.1).

2.1.1. *Classification of RFID systems*

Because of the variety of devices, features, applications and uses, there are a multitude of possible classifications due to the avaliability of a variety of devices, features and applications. However, we can retain some of the recurrent criteria found in these systems, such as:

– operating frequencies;

– types of transponders;

– modes of energy and data transmission;

– features.

Chapter written by Simon ELRHARBI and Stefan BARBU.

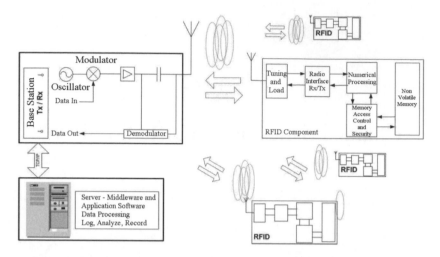

Figure 2.1. *Different elements of an RFID system*

2.1.2. *Available operating frequency ranges*

As RFID systems use electromagnetic waves, they have to comply with a number of regulations that can be applied at a national, regional or international level. In addition to local jurisdictions, RFID systems are subject to a number of works of standard committees which define radio frequency emission standards in terms of available radio frequencies, power levels and associated bandwidths. The available frequency ranges are of the ISM (Industrial-Scientific-Medical) type. RFID systems use the ISM frequencies and can be classified into four categories (Figure 2.2 and 2.3):

– low frequencies or LF (frequencies below 135 kHz);

– radio frequencies or HF (frequencies around 13,56 MHz);

– ultra-high frequencies or UHF (frequencies around 434 MHz, 869 - 915 MHz and 2,45 GHz);

– microwave or SHF (frequencies around 2,45 GHz).

2.1.3. *Transponder types*

In general, we distinguish between memory transponders and the microprocessor transponders (see sections 2.3.1 and 2.3.2 for more details). All transponders contain non-volatile memories (NVM). The basic operations on data in memory and microprocessor transponders are inputs/outputs, reading/writing operations in NVM memories and executions of cryptographic functions. Output operations deliver data outside and writing operations modify contents of an NVM memory. Therefore they are

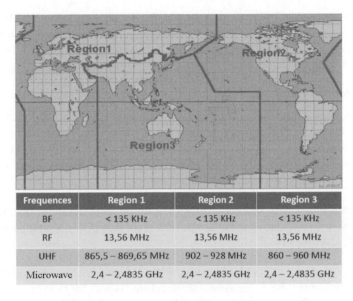

Frequences	Region 1	Region 2	Region 3
BF	< 135 KHz	< 135 KHz	< 135 KHz
RF	13,56 MHz	13,56 MHz	13,56 MHz
UHF	865,5 – 869,65 MHz	902 – 928 MHz	860 – 960 MHz
Microwave	2,4 – 2,4835 GHz	2,4 – 2,4835 GHz	2,4 – 2,4835 GHz

Figure 2.2. *Allocation of ISM frequencies for RFID*

a. LF Band (119–135 kHz)

USA/Canada	Europe	Japan	China
2400/f(inkHz)µV/m @ 300m	119 – 127 kHz: 66 dBµA/m @ 10 m 127 – 135 kHz: 42 dBµA/m @ 10 m	30 V/m @ 3m	P_{peak} < 1W

b. HF Band (13.56 MHz)

USA/Canada	Europe	Japan	China
13.553–13.567 MHz 42 dBµA/m @ 10 m	13.553–13.567 MHz 42 dBµA/m @ 10 m	13.553–13.567 MHz 42 dBµA/m @ 10 m	13.553–13.567 MHz 42 dBµA/m @ 10 m

c. UHF Band (860-960 MHz)

USA/Canada	Europe[1]	Japan	China
902 – 928 MHz $P_{e.i.r.p.}$[2] = 4W	865.0 – 868.0 MHz $P_{e.r.p.}$ = +20 dBm 865.6 – 868.0 MHz $P_{e.r.p.}$ = +27 dBm 865.6 – 867.6 MHz $P_{e.r.p.}$ = +33 dBm	952 – 955 MHz $P_{e.r.p.}$ = 1 W +6dB antenna gain = 4 W	840.5 – 844.5 MHz $P_{e.r.p.}$ = 2 W 920.5–924.5 MHz $P_{e.r.p.}$ = 2 W (Available since May 2007)

d. Microwave Band (2.45 GHz)

USA/Canada	Europe	Japan	China
2.400 – 2.483 GHz $P_{e.i.r.p.}$ = 4W	2.446 – 2.454 GHz $P_{e.i.r.p.}$ = 500 mW or 4 W (indoors)	2.400 – 2.4835 GHz 3 mW/MHz(Pe.i.r.p. = 1 W)	2.400 – 2.425 GHz 250 mW/m @ 3 m (Pe.i.r.p. = 21 mW)

[1] Listen-before-talk for 200 kHz channels. [2] Equivalent isotropically radiated power (e.i.r.p) = 1.64 × Effective Radiated Power (e.r.p.)

Figure 2.3. *Regulation of ITU frequencies for RFID*

particularly sensitive. This is why a logic security block is added to NVM memories, and is interposed between the NVM block and the input/output block. The role of the logic security block is to connect each output or writing operations to the hardware security of a chip.

2.1.3.1. *Memory transponders*

NVM memory cards have simple protection features for writing operations. Such protection features can make writing operations impossible. Some chips have their NVM memories associated with the programmable logic devices (PLD) which perform simple operations such as the verification of codes, the control of reading/writing operations, frozen cryptographic algorithms, etc.

Memory transponders, in addition to NVM memories, also contain memories which are engraved and fixed during manufacture (ROM for Read Only Memory) and volatile memory (RAM memory).

There are two types of NVM memories: EEPROM and FLASH. Electrically erasable and programmable, their fundamental structure is of MOS-type (metal oxide semiconductor). During programming, they use a Fowler-Nordheim Tunneling mechanism which deforms the potential barrier under a relatively high electric field. Carriers (typically electrons) are injected from the metal into the semiconductor through the oxide, resulting in deformation of its barrier potential. This mechanism causes some degradation both at the MOS interface level and in the oxide itself. The global electrical characteristics of memory logic gates can be altered. These memories therefore have two drawbacks:

– the endurance and the retention time of information that can be altered over time, which can reduce the life duration of these cards if there are too many writing cycles (typically > 100,000 cycles);

– the access to writing and the latency due to the electron injection mechanisms in the conduction band of the semiconductor.

Research toward higher integration densities, new architectures and robustness in CMOS memory continues to provide results, particularly in the design of circuits with very low power consumption which improves the memory endurance and the logic data processing time. However, the access to writing operations tends to be limited due to the programming mechanism. New memory technologies, called FeRAM, based on ferromagnetic properties of materials consisting of PZT (Pbx Zr1-x Ti O3) layers and more recently the SBT (Sr Bi2 Ta2 O9), significantly enhance the endurance and latency (<100 ns) properties with low power consumption. They represent an interesting alternative for smart cards and RFID chips which are designed and manufactured with the well established silicon technology [PEA 07]. The problems related to the development of manufacturing process, the integration in volume and the robustness, from both industrial and functional points of view, are yet to be solved since in the latter case, existing reading conditions can be destructive. Other memory technologies such as MRAM also offer promising alternatives, even when they are compared to FeRAM technologies. They operate through the magnetic polarization of materials. These materials constitute artificial structures whose composition includes alternately placed magnetic metal and a non-magentic material. These memories have magneto-resistive properties (i.e. cause a significant resistance change under the effect of an external magnetic field). These MRAM technologies provide an unlimited endurance and non-destructive readings contrary to FeRAM technologies.

The new FeRAM and MRAM memories combine features and advantages of SRAM, DRAM, EEPROM, FLASH and allow simplification of RFID chip architectures.

2.1.3.2. *Microprocessor transponders*

By the early 1980s, the main integration problems of microprocessors and NVM memories on a single CMOS support for mass production were highlighted. The deployment of smart cards was then naturally accompanied with the global development of CMOS circuits. Microprocessor cards have higher resources and are more flexible in terms of computing power as compared to the wired logic memories. They can trigger internal routines to protect data and control all logical and electrical signals to the non-volatile memory (in particular reading/writing operations). The main quality of a secure chip is that it can resist attacks, where the goal is to read confidential data stored in the NVM. The size of the chip remains unchanged (25 mm^2), since it is limited by mechanical constraint of flexion induced by the PVC support on the one hand and on the other hand, by the limited size due to the balance between the physical security and the complexity of the component.

The microprocessor chip design has to resolve a number of hardware challenges in terms of silicon technology: new materials, electrical and physical modeling, new architectures integrating more and more complex security sensors, innovative memories, cryptographic processors and communication interface, which is continually enriched and renewed (ISO-7816, ISO-14443, USB, NFC, etc.). Nevertheless, the power consumption issue remains unanswered since it increases with the rate of logic gate activities (in CMOS technology, the dynamic power consumption only occurs in each state transition of an elementary cell (inverter gate) and the rest of the electrical consumption, so-called static, comes essentially from leakage currents). In CMOS circuitry of microprocessors, each technological advance corresponds to an increase in frequency of about 43%, whereas the total capacity and the supply voltages are reduced by 30% and electric power is reduced by 50%. In contrast, the transistor density doubles after each generation, thus increasing the power density. In consequence, the density of the supply current is significantly increased. Consumption should be estimated and optimized at all levels, particularly at the system level, by determining the execution speed of tasks at all times of execution corresponding to an optimal schedule. It is necessary to implement a smart energy management and a task scheduling policy, as close as possible to the operating system which controls the architecture of all the modules of the chip. Rather than simply regulating the consumed electric power, the smart energy management must distribute various electrical quantities (current, voltage and electrical charge) as a function of the activity time of various modules, in particular when there are consumption peaks or standby modes. In addition, the management of the dissipated power should be considered and integrated during the design and manufacturing of micro-modules protecting transponders.

The CMOS circuit design was initially based on the paradigm of global synchronization where the center element is a single clock source. However, in recent years, we have developed a significant number of synchronous circuits due to an interest in this kind of RFID technologies. The formalism of asynchronous circuits is not based on a global synchronization but on local synchronization between different logic blocks. This leads to the development of a communication protocol, called *Handshake*, between logic blocks which allows an exchange of valid data using requests and acknowledgements. An advantage of this asynchronous technology lies in low power consumption since each unsolicited element will wait for a set of valid inputs. This activity is thus reduced to a minimum. Incidentally, it results in low electromagnetic emissions since the local operations are performed randomly in time, thus making a power consumption analysis attack (so-called side-channel attack) more difficult [CAU 05].

2.1.4. *Energy and data transmission modes*

2.1.4.1. *Energy transmission modes*

An analysis of wavelengths that are associated with operating frequencies used in ISM bands (typically about 3,000 to 3 meters for LF and HF frequencies and up to ten centimeters for UHF), and comparing the wavelengths with common antenna dimensions (typically several centimeters to ten centimeters) shows that there are only two kinds of electromagnetic waves:

– in RFID systems where the wavelength is much larger than the size of the antennae, the nature of electromagnetic waves is inductive. When the antennae of the reader are near the RFID chip, the power transmission is performed by mutual induction, where the electromagnetic field generated by the reader behaves as a primary circuit, and induces an electromotive force into the RFID chip antenna which behaves as a secondary circuit;

– in RFID systems where the wavelength is of the same order as the size of antennae, the nature of electromagnetic waves is radiative. In this case, energy is propagated by the source constituted by the reader.

The radio emissions with maximum power are regulated, notably in Europe by the European Radiocommunications Committee (ERC) in collaboration with the European

Telecommunications Standards Institute (ETSI). The commonly transmitted powers are around several hundred mW and are based on the application context, the limitations of power blocks, the configurations of antennae and their electromagnetic environment.

The electric power required for the RFID tags varies between ten μW and one hundred mW, depending on the number of logic gates, their activities and the general architecture of the circuit.

Depending on the application context and available powers, the transmission modes can be:

– passive: RFID chip is powered by the electromagnetic wave emitted by the reader;

– semi-passive: the RFID chip is assisted by a battery to power its circuitry and improve, to some extent, the sensitivity of the radio signal reception block. Indeed, the key factors that limit the zone of communication between the RFID tags and the readers are energy collection and radio sensitivity. The chip design is generally based on the balance between maximizing the energy collection and maximizing the sensitivity of the radio reception block;

– active: an active RFID chip allows an energetic retro-reflection by transmitting power rather than a simple retro-reflection of the transmitted wave, where only a part of the wave is captured by the chip. In other words, the active battery compensates for path loss. The inclusion of an energy source increases the operating zone, including the zone in an unfavorable electromagnetic environment since it can provide a signal strength gain of about 20 dBm as compared to passive RFID tags. However, interferences due to the retro-reflection energy sources appear and the multiplication of radio echoes renders the localization of chips by readers more difficult. Protocols and algorithms which can partially correct multipath interference impacts can be incorporated in active propagation of UHF. However these protocols increase the power consumption due to the activity of logic gates contained in the CMOS chip circuitry. The life span of embedded energy sources is thus altered and the application field is limited.

2.1.4.2. *Data transmission modes*

Most of the RFID tags are passive. The magnetic field is predominant in the LF and HF range of frequencies. When RFID chips are placed in an emitted magnetic field, they are viewed as loads by the reader due to mutual induction phenomena and variations in electrical quantities which they generate. To transmit data, when energy is adequate, the RFID chip modulates its load, which in turn modulates the magnetic field emitted by the reader. The nature of the load can be resistive or capacitive. Most of the RFID chips use the modulation mechanism of a resistive load (Figure 2.4).

In UHF ranges, the coupling effect is electromagnetic. When the incident wave emitted by the reader encounters an RFID obstacle, it is reflected. In both conditions

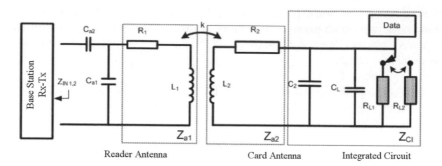

Reader Antenna   Card Antenna   Integrated Circuit

Figure 2.4. *Inductive coupling and load modulation*

with adequate incident power and taking into account the path loss, the RFID chip modulates its impedance according to the data rhythm which it wants to transmit. The variation of the chip impedance can be either resistive, capacitive, or both. Due to the ratio between the absorbed energy and the reflected energy, information is then transmitted by modulating the reflected electromagnetic field (Figure 2.5).

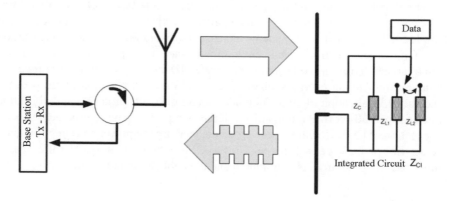

Figure 2.5. *Electromagnetic coupling and impedance modulation*

2.1.4.3. *Data and energy transfer procedures*

We can classify RFID systems based on size and quantity of information passing between different components of the RFID system. The information size can be:

− 1 bit. The principle is based on a binary detection (absence or presence) of the RFID tag in the interrogation zone of the reader with an option to disable (usually by physical destruction of an element of the tag). The types of applications include Electronic Article Surveillance (EAS). For LF or HF frequency ranges, where the mutual induction mechanism is preponderant, the reader generates a magnetic field with a frequency sweep. In this case the tags usually consist of a simple LC resonant circuit. At resonance, there is an oscillation causing an amplitude change at the reader. The oscillation depends on the distance between the chip and the reader and the quality factor of the antennae (Figure 2.6).

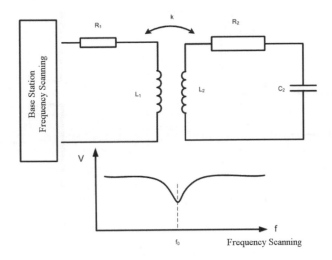

Figure 2.6. *Inductive coupling and resonant circuit*

Also at this frequency range, certain EAS systems are based on frequency division to be detected, while certain other EAS systems use materials based on ferromagnetic alloys having magnetic hysteresis properties.

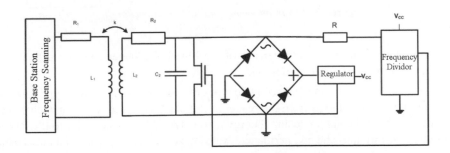

Figure 2.7. *Inductive coupling and frequency division*

For higher frequency ranges, we use RFID components with non-linear electrical characteristics (typically a varactor diode). These components are connected to an antenna adjusted to a UHF operating frequency. At resonance, the varactor diode generates current and retransmits a signal containing harmonics of the incident signal. The reader is then responsible for detecting the second harmonic of the reflected signal (Figure 2.8).

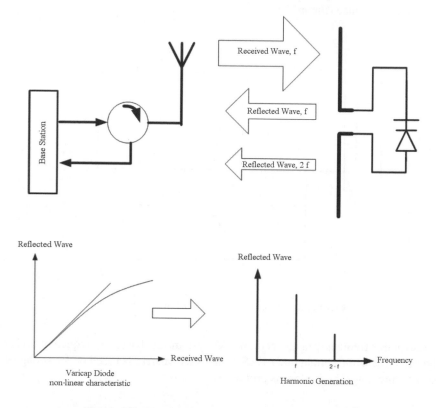

Figure 2.8. *Electromagnetic coupling and varactor diode*

Note that the number and intensity of the harmonics depend on the profile (or gradient) of the varactor diode doping.

– n bits. We differentiate between two types of procedures. In the first type, energy is continuously transferred from the reader to the chip, irrespective of the exchanged data flow. In the second procedure, energy is transferred from the reader to the RFID chip in a synchronous and non permanent way, i.e. sequential with the exchanged data flow.

For RFID systems performing a continuous energy transfer during communication, we can refine the classification by considering the manner in which the data exchange is performed. The data exchange may be simultaneous or alternate between the reader and the RFID chip.

The main energy and data transfer modes are:

– *Full Duplex* or FDX. Energy is continuously transferred during communication. Data are simultaneously exchanged between the RFID chip and the reader (Figure 2.9). For LF and HF ranges, energy is transmitted by inductive coupling and data frames are exchanged irrespective of their status during reception. As the reader and the RFID chip function as both receivers and data transmitters, it is necessary to separate two simultaneous modulations on the same carrier signal. In general, the RFID chip creates a sub-carrier to generate a frequency shift between return channel communications and forward channel communications. For higher frequency ranges, because of the presence of incident and reflected waves, the discrimination between the forward signal and the return signal, at the reader level, is provided by a directional coupler,

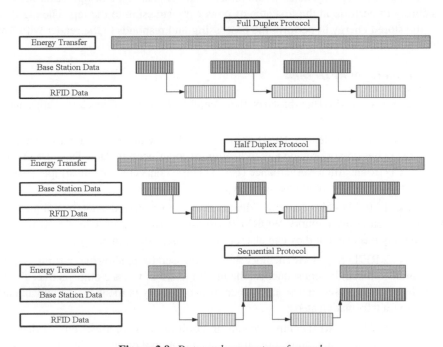

Figure 2.9. *Data and energy transfer modes*

– *Half Duplex* or HDX. Energy is continuously transferred during communication. Data is alternately exchanged between the RFID chip and the reader (Figure 2.9). In this case, the modulation of the return channel carrier is performed according to the change in the modulation of the forward channel. The reader and the RFID chip then successively function as transmitter and receiver until the end of communications, determined by the application. This HDX communication type ensures simplicity in the electronic architectures, although such architectures can be more expensive in terms of production since some chips need to have external capacities to store energy while answering.

– *Sequential* or SEQ. Energy transfer is not continuous, but occurs at regular intervals. Energy is transferred while data is sent from the reader to the RFID chip (Figure 2.9). The carrier is modulated by the reader during data transmission. When more data is to be transmitted, the carrier sent to the antenna is stopped. The main advantage of this procedure for data and energy transfer is that tags are not required to receive fixed carriers. Phase and frequency may be freely allocated. A modulation on these two parameters is performed. However, the added flexibility at the modulation level of the return channel leads to an extinction of the remote supply of the tag that will be mitigated by adding capacitors or batteries. This protocol is often used when communications are only done on the return channel. This type of operation offers a possibility to perform in the first step of power transmission to the tag. The tag then uses the stored energy to perform the processing and responds to the reader with low HF emissions [CAU 05].

2.1.5. *Features of RFID chips*

We can classify RFID tags based on their features [ENG 03]. Five classes are usually defined:

– class 0: in this category, RFID chips have a binary detection function (absence or presence) but do not provide any indication of identification data. They are mainly used for electronic article surveillance (EAS). These chips can contain simple passive components without wired logic circuitry;

– class 1: RFID chips have an identification function and contain unique data stored in a read-only memory WORM (*write-once read-many*). Class 1 RFID tags are generally passive, but they can also be semi-passive or even active;

– class 2: RFID chips have memory that can be used for reading and writing, where data can be updated. By consequence, the main feature of Class 2 chips is traceability since they can be reused in a process where it is necessary to identify and update data related to transformation steps;

– class 3: Class 3 RFID chips contain, in addition to memories, sensors that can record data such as temperature, pressure or acceleration. For processing data, sensors require a storage element even in the absence of RFID readers. Energy source should be embedded in the RFID chips for semi-passive or active use;

– class 4: RFID chips, equipped with sensors and memories, have adequate resources to process data and communicate independently to establish wireless ad hoc networks between them. They should be active to initiate communication. From a functional point of view, these chips are part of the "smart dust" paradigm developed by researchers at the University of Berkeley in California. Their goal is the deployment of distributed networks with thousands of nodes that are Class 4 RFID chips.

2.2. Transmission channel

RFID systems use a electromagnetic transmission channel (free propagation medium) which will be described using Maxwell's equations.

2.2.1. *Maxwell's equations*

Maxwell's equations describe the behavior of the electromagnetic field which is produced by distribution of electric charge in the space, where the distribution varies with time. Their local expressions are:

$$\nabla \times \vec{E} = -\mu \cdot \frac{\partial \vec{H}}{\partial t} \text{(Maxwell-Faraday)}$$

$$\nabla \times \vec{H} = \vec{J} + \epsilon \cdot \frac{\partial \vec{E}}{\partial t} \text{ (Maxwell-Ampère)}$$

$$\nabla \cdot \vec{E} = \frac{\rho}{\epsilon} \text{ (Maxwell-Gauss)}$$

$$\nabla \cdot \vec{H} = 0 \text{ (Conservation of magnetic flux)}$$

where:

$\vec{E}$: electric field(V/m)

$\vec{H}$: magnetic field (A/m)

$\vec{J}$: electric current density (A/m^2)

ρ: electric charge density (C/m^3)

μ: magnetic permeability of the medium (H/m)

ϵ: electric permeability of the medium (F/m)

These equations are formalized by constitutive relations, which describe the behavior of fields inside materials:

$$\vec{D} = \epsilon \cdot \vec{E}$$

$$\vec{B} = \mu \cdot \vec{H}$$

$$\vec{J} = \sigma \cdot \vec{E}$$

with:

$\vec{D}$: electric induction (C/m^2)

$\vec{B}$: magnetic induction (T)

σ: conductivity of the material (S/m)

The propagation medium is supposed to be isotropic and consequently the permeability μ and ϵ are considered as scalar quantities.

In harmonic regime, at the frequency ω, $\frac{\partial}{\partial t} = -j \cdot \omega$ and the equations are:

$$\nabla \times \vec{E} = j \cdot \omega \cdot \mu \cdot \vec{H}$$
$$\nabla \times \vec{H} = \vec{J} - j \cdot \omega \cdot \epsilon \cdot \vec{E}$$

These Maxwell relationships will enable us to determine the configuration of electric and magnetic fields for RFID systems.

2.2.2. *Electromagnetic field generated by an electric dipole*

Consider an electric dipole (also called Hertzian dipole) consisting of a current segment of elementary length dl flowed through by a current I which assumed to be uniform (see Figure 2.10).

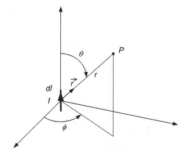

Figure 2.10. *Ideal dipole*

The field radiated by the electric source of the dipole is calculated, in a classical manner, through the potential vectors and scalars created at any point in the space by a distribution of charges and currents.

β is the number of waves defined by the relation: $\beta = \frac{2 \cdot \pi}{\lambda} = \frac{\omega}{C}$ and η_0 impedance of vacuum defined by: $\eta_0 = \sqrt{\frac{\mu_0}{\epsilon_0}} = 120 \cdot \pi$, where β and η_0 are given.

The components of the field generated by the electric dipole are given in spherical coordinates by the following relations:

$$\vec{E}_r = -\frac{I \cdot dl}{4 \cdot \pi} \cdot \eta_0 \cdot \beta^2 \cdot 2 \cdot cos\theta \cdot \left[\frac{1}{(j \cdot \beta \cdot r)^2} + \frac{1}{(j \cdot \beta \cdot r)^3} \right] \cdot e^{-j \cdot \beta \cdot r} \cdot \vec{u}_r$$

$$\vec{E}_\theta = -\frac{I \cdot dl}{4 \cdot \pi} \cdot \eta_0 \cdot \beta^2 \cdot sin\theta \cdot \left[\frac{1}{(j \cdot \beta \cdot r)} + \frac{1}{(j \cdot \beta \cdot r)^2} + \frac{1}{(j \cdot \beta \cdot r)^3} \right] \cdot e^{-j \cdot \beta \cdot r} \cdot \vec{u}_\theta$$

$$\vec{H}_{\varphi} = -\frac{I \cdot dl}{4 \cdot \pi} \cdot \beta^2 \cdot \sin\theta \cdot \left[\frac{1}{(j \cdot \beta \cdot r)} + \frac{1}{(j \cdot \beta \cdot r)^2} \right] \cdot e^{-j \cdot \beta \cdot r} \cdot \vec{u}_{\varphi}$$

where $\vec{u}_r$, $\vec{u}_\theta$ and $\vec{u}_\varphi$ are direction vectors of the spherical reference.

2.2.3. Electromagnetic field generated by a magnetic dipole

Consider a magnetic dipole which consists of a current loop, where the perimeter of the loop is less than a quarter of the wavelength of the power source I that is assumed to be uniform (Figure 2.11).

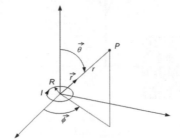

Figure 2.11. *Magnetic dipole*

The equations of the field radiated by the electric dipole and the equations of the field radiated by a magnetic dipole have a dual nature. We move on from an expression of the field emitted by the electric dipole to that of a magnetic dipole by replacing E by H and the current element Idl by the magnetic moment given by $I \cdot \pi \cdot R^2$ where I is the current flowing through a elementary coil of radius R.

We obtain the components of the field generated by the magnetic dipole in spherical coordinates:

$$\vec{H}_r = -\frac{\beta^2 \cdot \pi \cdot R^2 \cdot I \cdot \cos\theta}{2 \cdot \pi \cdot r} \cdot \left[\frac{1}{(j \cdot \beta \cdot r)} + \frac{1}{(j \cdot \beta \cdot r)^2} \right] \cdot e^{-j \cdot \beta \cdot r} \cdot \vec{u}_r$$

$$\vec{H}_\theta = -\frac{j \cdot \beta^3 \cdot \pi \cdot R^2 \cdot I \cdot \sin\theta}{4 \cdot \pi} \cdot \left[\frac{1}{(j \cdot \beta \cdot r)} + \frac{1}{(j \cdot \beta \cdot r)^2} + \frac{1}{(j \cdot \beta \cdot r)^3} \right]$$
$$\cdot e^{-j \cdot \beta \cdot r} \cdot \vec{u}_\theta$$

$$\vec{E}_\varphi = j \frac{\eta_0 \cdot \beta^3 \cdot \pi \cdot R^2 \cdot I \cdot \sin\theta}{4 \cdot \pi} \cdot \left[\frac{1}{(j \cdot \beta \cdot r)} + \frac{1}{(j \cdot \beta \cdot r)^2} \right] \cdot e^{-j \cdot \beta \cdot r} \cdot \vec{u}_\varphi$$

2.2.4. *Field zones surrounding antennae*

The space surrounding the transmitting antenna of the reader can be divided into three zones (Figure 2.12):

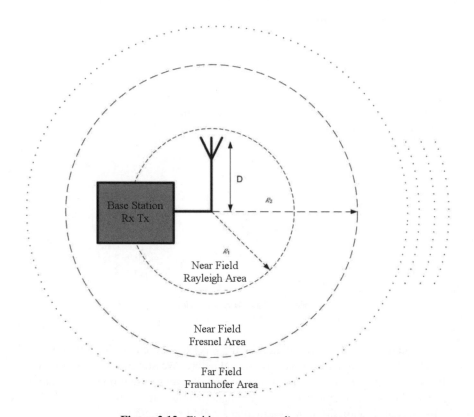

Figure 2.12. *Field zones surrounding an antenna*

– the zone closest to the antenna, where $\beta \cdot r << 1$ is the Rayleigh zone or "near field" zone. Its boundary is represented by the radius R_1 which varies according to the relation:

$$R_1 \leq 0,62 \cdot \sqrt{\frac{D^3}{\lambda}}$$

– the remote zone from the antenna, where $\beta \cdot r >> 1$ is the Fraunhofer zone or "far field" zone. Its boundary is represented by the radius R which varies according to the relation:

$$R \geq \frac{2 \cdot D^2}{\lambda}$$

– a Fresnel zone is an intermediate "near field" zone. The limits of boundaries correspond to the following equations: $0,62 \cdot \sqrt{\frac{D^3}{\lambda}} \leq R_2 \leq \frac{2 \cdot D^2}{\lambda}$. The transition zone between the near and far field depends on the geometry of the antenna.

2.2.4.1. Near field zones

In the case of electric dipoles, in near fields, where $\beta \cdot r << 1$, the third-order term of the electromagnetic field expression becomes predominant and $e^{-j \cdot \beta \cdot r}$ tends to 1. Components of the field become:

$$\vec{E}_{r,NF} = -j \cdot \frac{I \cdot l \cdot \eta_0 \cdot cos\theta}{2 \cdot \pi \cdot \beta \cdot r^3} \cdot \vec{u}_r$$

$$\vec{E}_{\theta,NF} = -j \cdot \frac{I \cdot l \cdot \eta_0 \cdot sin\theta}{4 \cdot \pi \cdot \beta \cdot r^3} \cdot \vec{u}_\theta$$

$$\vec{H}_{\varphi,NF} = \frac{I \cdot l \cdot sin\theta}{4 \cdot \pi \cdot r^2} \cdot \vec{u}_\varphi$$

The index NF refers to near fields.

In the case of dipole magnetic, components of the field become:

$$\vec{H}_{r,NF} = \frac{\pi \cdot R^2 \cdot I \cdot cos\theta}{2 \cdot \pi \cdot r^3} \cdot \vec{u}_r$$

$$\vec{H}_{\theta,NF} = \frac{\pi \cdot R^2 \cdot I \cdot sin\theta}{4 \cdot \pi \cdot r^3} \cdot \vec{u}_\theta$$

$$\vec{E}_{\varphi,NF} = -j \cdot \frac{\eta_0 \cdot \beta \cdot \pi \cdot R^2 \cdot I \cdot sin\theta}{4 \cdot \pi \cdot r^2} \cdot \vec{u}_\varphi$$

In near fields, we observe a phase shift of 90 degrees between $\vec{E}_{NF}$ and $\vec{H}_{NF}$, which indicates an energy exchange of reactive nature (where the stored energy is much larger than the radiated energy) between the antenna and the external environment. Moreover, $\vec{E}_{NF}$ and $\vec{H}_{NF}$ do not vary in the same way:

– for electric dipoles, the electric field is predominant in the plane $(\vec{\theta}, \vec{r})$, it varies in $\frac{1}{r^3}$ and behaves like a static electric dipole. The magnetic field is negligible in the transverse plane (it varies in $\frac{1}{r^2}$). The behavior of electric dipoles in the near field zone is of minor importance except for minority technologies for specific applications that use electrical coupling for passive transponders;

– for magnetic dipoles, the magnetic field is dominant in the plane $(\vec{\theta}, \vec{r})$, and the electrical component is negligible in the transverse plane. This property of the electromagnetic field will be used as advantages, particularly in the energy transmission mechanism by mutual induction.

2.2.4.2. *Far field zones*

In the case of electric dipoles, in far fields, where $\beta \cdot r \gg 1$, the first-order term of the electromagnetic field expression is predominant. The components of the field are:

$$\vec{E}_{\theta,FF} = j \cdot \frac{I \cdot l \cdot \eta_0 \cdot sin\theta \cdot \beta}{4 \cdot \pi \cdot r} \cdot e^{-j\cdot\beta\cdot\, r} \cdot \vec{u}_\theta$$

$$\vec{H}_{\varphi,FF} = j \cdot \frac{I \cdot l \cdot \beta \cdot sin\theta}{4 \cdot \pi \cdot r} \cdot e^{-j\cdot\beta\cdot\, r} \cdot \vec{u}_\varphi$$

The index $_{FF=}$ refers to far fields.

In the case of magnetic dipoles, in far fields, the components of the field are:

$$\vec{H}_{\theta,FF} = -\frac{\beta^2 \cdot \pi \cdot R^2 \cdot I \cdot sin\theta}{4 \cdot \pi \cdot r} \cdot e^{-j\cdot\beta\cdot\, r} \cdot \vec{u}_\theta$$

$$\vec{E}_{\varphi,FF} = \frac{\eta_0 \cdot \beta^2 \cdot \pi \cdot R^2 \cdot I \cdot sin\theta}{4 \cdot \pi \cdot r} \cdot e^{-j\cdot\beta\cdot\, r} \cdot \vec{u}_\varphi$$

In far fields, $\vec{E}_{FF}$ and $\vec{H}_{FF}$ represent the typical characteristics of a plane wave:
– they are in phase;
– both of them vary in $\frac{1}{r}$;
– they are mutually orthogonal;
– the plane $(\vec{E}_{FF}, \vec{H}_{FF})$ is perpendicular to the direction of propagation.

2.2.5. *Wave impedance*

The ratio $\frac{E}{H}$ is homogeneous for an impedance defined as the wave impedance. The impedance varies depending on near or far field zones. However this impedance, in case of electric or magnetic dipole, tends to a limit in far field, which is one of the impedances of the wave in free space, previously defined by: $\eta_0 = \sqrt{\frac{\mu_0}{\epsilon_0}} = 120 \cdot \pi = 377$ ohms (Figure 2.13).

In near fields in the Rayleigh zone:
– electric dipoles present a wave impedance η which is higher than that of vacuum and varies as: $\eta = \frac{\eta_0}{\beta \cdot r}$,
– magnetic dipoles, on the contrary, present a wave impedance η which is smaller than that of vacuum and varies as: $\eta = \eta_0 \cdot \beta \cdot r$.

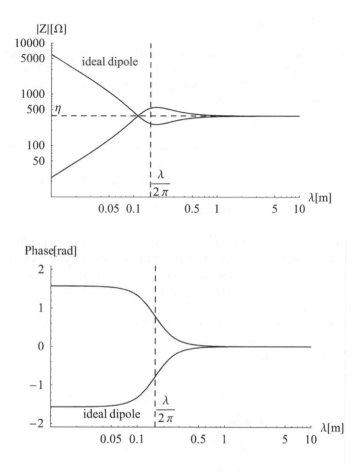

Figure 2.13. *Wave impedance (module and phase) in function of the distance from antenna*

In far fields, in the Fraunhofer zone, the wave impedances of electric and magnetic dipoles converge to a limit that is the impedance of vacuum η_0.

The border between the Rayleigh zone and the Fraunhofer zone is of the order of $\frac{\lambda}{2 \cdot \pi}$ and corresponds to the previously observed transition of the Fresnel zone.

The phase of the wave impedance shows the reactive nature of near fields and the radiative nature of far fields. Moreover, in the near-field zone, we can note that the phase of the wave impedance for magnetic dipoles is positive, which characterizes the predominance of the inductive coupling which we want to maximize. On the other hand, the wave impedance of electric dipoles characterize a capacitive coupling which we want to minimize except in the case of specific and unconventional applications and technologies.

2.2.6. *Antenna impedance*

Antennae, wired or in loops, are the seat of electromagnetic perturbation and radiate energy that can be reactive or radiative. They constitute a transition device, i.e. impedance transformer, between the guided propagation medium and the free propagation medium. In order that this transition is performed with maximum effect, an adjustment of the two media is necessary.

The adjustment quality of the antenna, depending on its geometry and supply mode, can be represented by its input impedance Z_{IN}. It is a combination of reactances representing stored energy and resistances, where resistances represent losses.

$$Z_{IN} = R_{radiative} + R_{ohmic} + j \cdot X$$

The ohmic resistances are essentially due to the phenomena of skin effect. Indeed, in high frequencies, electrical signals do not propagate in a homogeneous way throughout the transversal section of the conductor. They are distributed on the surface of the conductor (Figure 2.14). The electric current density decreases exponentially from the surface to the interior of the conductor. The skin effect is represented by the thickness δ that is proportional to $\frac{1}{\sqrt{f}}$ where f is the frequency of the electrical signal. The higher the frequency, the lower the efficiency will be and the higher the resistance of the section of the conductor will be.

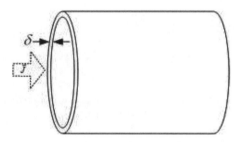

Figure 2.14. *Skin effect for a cylindrical conductor ($\vec{J}$ represents the current density)*

Radiative resistances characterize the losses due to radiation and are determined by the geometry of antennae and the wavelengths of electromagnetic signals.

2.2.7. *Radiated power*

The Poynting relation, where $\vec{P} = \frac{1}{2} \cdot (\vec{E} \otimes \vec{H}^*)$, indicates the direction of propagation. The module of the Poynting vector gives us the radiated power density P_t by the following relations:

$$p = \left| \vec{P} \right| = \frac{1}{2} \cdot \left| \vec{E} \right| \cdot \left| \vec{H} \right|$$

Given $p = 15 \cdot \pi \cdot \frac{I^2 \cdot dl^2 \cdot sin^2\theta}{\lambda^2 \cdot r2}$, then the radiated power is: $P_t = \int \int p \cdot dS$, with dS: surface element in spherical coordinates. We finally obtain:

$$P = 40 \cdot (\frac{\pi \cdot I \cdot dl}{\lambda})^2$$

The Poynting vector can also be written as $\widetilde{P} = P + j \cdot Q$ where P is the active power (or radiative) and Q the reactive power. In near fields, the reactive power is predominant and in far fields, the active power is dominant and the reactive power is zero.

Following the field zones surrounding the antennae, the radiated power density can be:

– constant in the Rayleigh zone (near fields);

– decreasing in the Fraunhofer zone (far fields);

– fluctuating in the Fresnel zone.

2.2.8. *Near-field coupling*

As discussed earlier, in near field zones, the electric field is decoupled from the magnetic fields where one field dominates the other depending on the nature of antennae and their configuration. In the case of current loops, the magnetic field is dominant as compared to the electric field while for the electric dipoles, the situation is reversed because of their dual nature. Hence, we will perform power transmission from a distance, i.e. by an electric coupling, using capacitors to interact with the electric field, or by a magnetic coupling which uses inductors to interact with the magnetic field.

RFID technologies very broadly use inductive coupling that we will explain in detail.

2.2.8.1. *Inductive coupling*

The energy stored in the magnetic field can be converted into electrical potential difference by electromagnetic induction: the flow of a magnetic field by the surface of a closed loop induces an electromotive force within the limits of the loop. We start from the Maxwell-Faraday equation:

$$\nabla \times \vec{E} = -\mu \cdot \frac{\partial \vec{H}}{\partial t} (\text{Maxwell-Faraday})$$

A temporal variable magnetic field generates an electromotive force:

$$e = -\frac{\partial \Phi}{\partial t}$$

where:

$$\Phi = -\mu \cdot \oint_S H \cdot dS$$

is the flux generated by the H-field crossing the surface S. The result is:

$$e = -\mu \cdot \frac{\partial}{\partial t} \left(\oint_S H \cdot dS \right)$$

As we have seen previously, a loop carrying an electric current generates a magnetic field. When second loop is introduced, this magnetic field induces an electromotive force in the second loop and thereby generates an electric current. This is the basic operating principle of RFID technologies in near fields, where ISM frequency ranges are lower than 135 KHz or equal to 13.56 MHz. Note that the induced current generates, in its turn, a magnetic field which opposes the inductor field (according to Lenz's law).

2.2.8.2. Proper inductance

In the case of two coupled magnetic loops, the field generated by the first circuit generates a magnetic flux in the second circuit and vice versa (Figure 2.15).

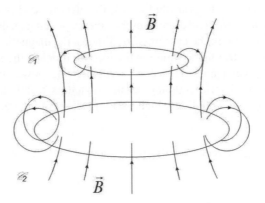

Figure 2.15. *Mutual inductance between two circuits C_1 and C_2*

Denoted as Φ_{11} and Φ_{22}, flows generated by two loops on their surface, Φ_{12} and Φ_{21}, are flows respectively generated by the circuit 1 flowed through by the current I_1 which generates a magnetic field H_1 on the circuit 2 and the circuit 2, flows where current I_2 and generates a magnetic field H_2 on the circuit 1.

The proper inductance of each loop is then respectively defined as:

$$L_1 = \frac{\Phi_{11}}{I_1} = \mu \cdot \frac{\oint_{S_1} H_1 \cdot dS_1}{I_1}$$

$$L_2 = \frac{\Phi_{22}}{I_2} = \mu \cdot \frac{\oint_{S_2} H_2 \cdot dS_2}{I_2}$$

Based on their geometry, we can calculate the inductance of the antennae from these equations. However, for complex forms, the analytical formulation can be difficult to determine. We provide semi-empirical expressions, in general, due to this reason.

We have presented above several semi-empirical formulas to calculate inductances. The inductance L is given by μH, dimensions in centimeters and L_0 represents a section of infinitesimal width.

For a circular coil of circular section (Figure 2.16):

$$L = 0.002 \cdot \pi \cdot D \cdot \left[ln \left(\frac{8 \cdot D}{d} \right) - 1.75 \right]$$

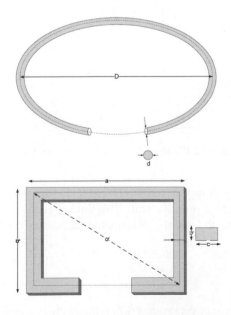

Figure 2.16. *Circular and rectangular coil*

For a rectangular coil of rectangular section (Figure 2.16):

$$L = 0.004 \cdot \left[a \cdot ln\left(\frac{2 \cdot a \cdot b}{c \cdot (a+d)}\right) + b \cdot ln\left(\frac{2 \cdot a \cdot b}{c \cdot (b+d)}\right) \right.$$
$$\left. +2 \cdot d - \frac{a+b}{2} + 0.447 \cdot (c+h) \right]$$

2.2.8.3. *Mutual inductance*

The mutual inductances of the circuit 1 on the circuit 2 and the of circuit 2 on the circuit 1 (Figure 2.17) are defined as:

$$M_{12} = \frac{\Phi_{12}}{I_1} = \mu \cdot \frac{\oint_{S_2} H_1 \cdot dS_2}{I_1}$$

$$M_{21} = \frac{\Phi_{21}}{I_2} = \mu \cdot \frac{\oint_{S_1} H_2 \cdot dS_1}{I_2}$$

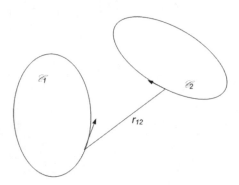

Figure 2.17. *Mutual inductance between two circuits*

The total magnetic flow Φ of a coil is linked to the potential vector A_1 located on an elementary portion of the second circuit. This linkage is due to current flowing through the first circuit, given: $\Phi = \oint_2 A_1 \cdot dl_2$, mutual inductance can be calculated by:

$$M_{12} = \frac{\mu_0}{4 \cdot \pi} \cdot \oint_{C_1} \oint_{C_2} \frac{dl_1 \cdot dl_2}{r_{12}} \text{(Neumann)}$$

In our case, because of the symmetry of influences, $\Phi_{12} = \Phi_{21}$ and by consequence $M_{12} = M_{21} = M$.

The mutual inductance calculation can be difficult since it must take into account the number of rounds of the coils, the geometry of the coils and their location in the space in mutual influence. The analytical formulation is not easy and often requires a semi-empirical approach.

2.2.8.4. *Coupling coefficient*

Rather than determining the mutual inductance, it may be easier to determine the common flux of two coils. This quantity is given by the coupling coefficient $k = \frac{\Phi_{12}}{\sqrt{\Phi_{11} \cdot \Phi_{22}}}$ or $k = \frac{M}{\sqrt{L_1 \cdot L_2}}$.

Due to mutual induction, the transponder is detected as a load at the reader level. It will cause a more or less mismatch, depending on the coupling, to the input of the reader antenna. Therefore, the determination of the coupling coefficient, which measures the amount of energy transmitted from the reader to the RFID chip, is an important step in the design and dimensioning of RFID and reader antennae.

2.3. First level electric model in inductive coupling

The couple reader/RFID transponder forms, by the configuration of antennae, two coupled resonant systems in which the resonant frequencies of each antenna are interdependent (Figure 2.18).

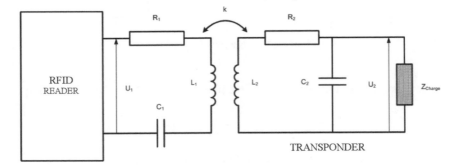

Figure 2.18. *First level electric model of the couple reader/RFID transponder*

At the reader (or base station) side, the resonant serial system (R_1, L_1, C_1) will create, at resonance, a maximum current i_1 which generates a magnetic field (and a negligible electric field). The magnetic field which is variable in time and space will induce an electromotive force in the antenna of the transponder. The induced current i_2 passing through the transponder antenna will, in turn, create a magnetic field which opposes the field of base station (Lenz's rule). The total field will be modified by the transponder load. The voltage across the chip, represented by a load impedance Z_{Charge} will be maximum at the resonance of parallel circuit (R_2, L_2, C_2).

The set can be modelled as a double-tuned transformer in which the primary represents the reader and the secondary represents the transponder with a coupling coefficient k.

2.3.1. *Magnetic loop*

The electrical model of a magnetic loop which generates a field or collects a magnetic flux can be represented by a resistance of ohmic losses in series with a self and a capacitor in parallel, parasite or not, which comes from the circuit connected with the antenna (Figure 2.19). If the antenna has several spires, then a capacitor is added in parallel. Losses due to radiation resistance are negligible at this frequency as compared to ohmic resistances.

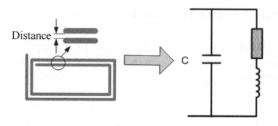

Figure 2.19. *Magnetic loop antenna*

The expression of the impedance of the magnetic loop (ML) Z_{ML} is given by:

$$\frac{1}{Z_{ML}} = \frac{1}{R + j \cdot \omega \cdot L} + j \cdot \omega C$$

The resonant frequency of the magnetic loop is derived by the calculation of module and phase of the impedance Z_{ML} which is 0. We obtain:

$$f_{BM,0} = \frac{1}{2 \cdot \pi} \cdot \sqrt{\frac{1}{L \cdot C} - \frac{R^2}{L^2}}$$

The resonant frequency calculation in RFID systems with inductive loop is not of much importance, since it depends on the mutual induction of other inductive circuits. During the antenna design, the presence of other cards in the field should be taken into account: if multiple cards are in the field, global mutual inductance increases, and the global resonance frequency decreases. In addition, the resonant frequency and the quality factor Q varies in function of the received power, and the activity of the chip.

2.3.2. Base station antenna

The base station (or RFID reader) sends power and data to the RFID chip. The performance of the base station, from an operational point of view depends on:

– the size of antennae;

– the adjustment of the antenna;

– the bandwidth and the quality factor of antennae;

– the emitted power;

– the influence of its electromagnetic environment.

The size of antennae is constrained by imposed by specifications and the application context besides purely aesthetic aspects which affect the nature of the materials, packaging and the form factors. The coupling that determines the amount of energy transmitted remotely from the reader to the RFID chip will also strongly depend on the antenna configuration.

For practical reasons, for example the interoperability of RFID electronic components, we may need to adjust the impedance of reader antenna to 50Ω to connect to the emission power stage of the base station. There are a multitude of antenna adjustment circuits. We may retain some typical impedance adjustment circuits, such as the one in a capacitive bridge (Figure 2.20).

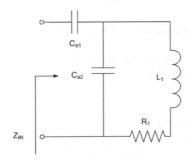

Figure 2.20. *Impedance adjustment in a capacitive bridge*

The input impedance Z_{IN} of the antenna is given by:

$$Z_{IN} = \frac{1}{j \cdot \omega \cdot C_1} + \frac{1}{j \cdot \omega \cdot C_2 + (R_1 + j \cdot \omega \cdot L_1)}$$

At the resonance of the circuit, we want to have $Z_{IN} = R_e = 50\Omega$. This corresponds to the following conditions for the adjustment capabilities C_{a1} and C_{a2}:

$$C_{a1} = \frac{\sqrt{R_1}}{\omega} \cdot \frac{1}{\sqrt{R_e \cdot (R_1^2 - R_1 \cdot R_e + \omega^2 \cdot L_1^2)}}$$

$$C_{a2} = (R_e - R_1) \cdot (\omega^2 \cdot L_1 \cdot R_e + \frac{R_1}{C_{a1}})$$

In some cases, we use antennae with symmetric structures where the antennae are connected to asymmetrical wires. We can also uses *balun* (short for balanced-unbalanced) assemblies that can overcome this problem by symmetrizing two equal currents in the antenna.

Once power is adjusted, the next step is to determine the over-voltage (quality) factor Q which fixes the amount of bandwidth to communicate with RFID chips. The quality factor must also transmit and receive data without signal deformation.

The quality of a resonant circuit is measured by the ratio between the useful stored energy and the energy lost during a period:

$$Q = 2 \cdot \pi \cdot \frac{\text{Stored energy}}{\text{Energy lost during a period}}$$

In order to ensure communication, the maximum quality factor Q_1 of the antenna should allow the bandwidth (in -3 dB) to pass the frequencies contained in the modulated signal, i.e. the carrier and the modulation sidebands. The bandwidth BW_1 is given by:

$$BW_1 = \frac{f_0}{Q_1}$$

where f_0 is the resonant frequency of the circuit.

The fundamental frequency of binary signals, where the signals are assumed to be square cycle and have ratio of 50 %, must be equal to half the bandwidth, i.e.

$$2 \cdot R = BW_{1,min}$$

where R is the binary throughput, hence:

$$Q_{1,max} = \frac{f_0}{2 \cdot R}$$

For example, for a carrier frequency at 13.56 MHz and a throughput of 106 kbit/s, we obtain a coefficient of 64. For a throughput of 848 kbit/s, the coefficient is only 8.

We should also take into account other parameters influencing the impulsive response of the antenna circuit, for example:

– the cut-off, of the carrier during breaks and during modulation at 100%;

– the rapid variation during a modulation at 10%.

Indeed, in transient mode, the transfer function of the circuit of a magnetic loop adapted in capacitive bridge has a simple form:

$$H(j \cdot \omega) = \frac{A_0 \cdot p}{\omega_0^2 + 2 \cdot \Delta \cdot p + p^2}$$

with:

$$p = j \cdot \omega$$

$$A_0 = \frac{C_{a1}}{L_1 \cdot (C_{a1} + C_{a2})}$$

$$\omega_0 = \frac{1}{\sqrt{L_1 \cdot (C_{a1} + C_{a2})}}$$

$$\Delta = \frac{R_1}{2 \cdot L_1}$$

The time constant θ of the circuit is given by:

$$\theta = \frac{1}{\Delta} = \frac{2 \cdot L_1}{R_1} = \frac{Q_1 \cdot T_0}{\pi}$$

where T_0 is the signal period.

When applying a voltage step to the circuit of transmitting antenna, its impulsive response $I(t)$ will oscillate until it reaches its final value:

$$I(t) = I_{max} \cdot e^{-\Delta \cdot t} \cdot cos\omega \cdot t$$

at $t = 3 \cdot \theta$, $I = 5\% I_{max}$. We can calculate the factor Q_1 of the antenna to pass an impulsion of width T_0.

The factor Q_1 is calculated. The next step is to design the RLC values of the antenna according to the output stage of the base station and its environment (Figure 2.21):

The quality factor of the antenna depends on R_1 as given by the relation $Q_1 = \frac{\omega \cdot L_1}{R_1}$. R_1 is obtained different contributions of resistive components of the antenna, the output resistance of the HF power stage connected to the antenna and losses due to Foucault leakage currents in the metal environment:

$$R_1 = R_{\text{output}} + R_{\text{L1, HF}} + R_{\text{series, 1}} + R_{\text{C1}}$$

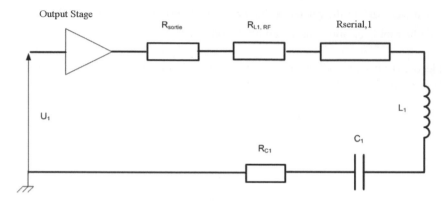

Figure 2.21. *Circuit of the base station antenna*

Resistive contributions are:

– R_{output} is the output resistance of the RF power stage connected to the antenna;

– $R_{\text{L1,HF}}$ is the resistive part of the losses due to Foucault current which can occur if the metal walls are near the base station antenna;

– $R_{\text{series,1}}$ is the series resistance of the basic antenna in which the skin effect should be included;

– R_{C1} is the parasitic resistance of the tuning capacitor of the antenna. Usually this resistance is negligible as compared to other resistive contributions of the antenna.

Once the adjustment is done to the characteristic impedance (here 50Ω), the over-voltage coefficient is set with the desired bandwidth. Then the next step is to determine the maximum current with flows through the antenna of transmitter base station in the magnetic field.

At resonance, the circuit of the magnetic loop behaves like a pure resistance and consumes power:

$$P_1 = R_1 \cdot I_1^2$$

that provides the power stage of the base station, where I_1 is the current flowing through the antenna. By introducing $Q_1 = \frac{\omega \cdot L_1}{R_1}$, we obtain the following relation:

$$I_1 = \sqrt{\frac{P_1 \cdot Q_1}{\omega \cdot L_1}}$$

The current I_1 should be controlled as a function of electrical parameters of the antenna and the power emitted by the power stage of the reader's output to adapt to specifications required by application, by the metal environment and radio standard constraints.

Regarding the metallic environment, the external magnetic field near a metal induces Foucault electric currents. These currents cause losses and an antenna mismatch, thus weakening the magnetic field which reduces the operational area of the couple reader/RFID card and increases the data transmission error rate. Generally, if possible, the metal walls must be outside the operational area. Otherwise, a ferrite protection can be incorporated. In all cases, antennae should always be readjusted (Figure 2.22).

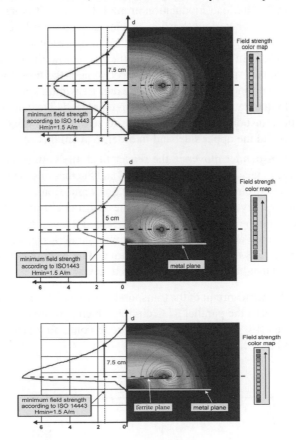

Figure 2.22. *Distribution of the magnetic field in vacuum in the presence of a metal wall and ferrite protection*

2.3.3. *RFID chip antenna*

The RFID chip receives power and data from the RFID reader. In the operational zone of the reader, we must consider the following constraints:

– The card must recover the energy emitted on a carrier frequency f_c by the RFID reader within the limits set by radio standards ETSI, FCC;

– When the energy is sufficient to power the RFID chip and the data transmitted by the reader is received and processed by the radio reception and the numerical blocks of the chip, the RFID chip will respond, according to its logic state diagram, by modulating a load at the rate of transmitted data. It creates a sub-multiple sub-carrier of f_c. At first glance, we can envisage to determine the electrical parameters of the transponder antenna to authorize the right transmission of data contained in one of the sidebands created by the load modulation around the carrier frequency. However, if several transponders, for various reasons, are juxtaposed, it induces a higher global mutual coupling, which generates, in turn, a drop in the resonant frequency resulting from all transponders. In some cases, it can then be below the carrier frequency or below the low sideband, perturbing the right reception of the transponder signal from the RFID reader. Thus one can receive only the data transmitted in the sideband higher than the carrier frequency. Also, it is generally recommended to calculate the electrical parameters of the transponder antenna which is to be tuned to a frequency higher than that of the carrier and the upper sideband in order to avoid this scenario;

– when the transponder moves in the reader field, the voltage across the antenna varies over a large range. To avoid over-tensions and fluctuations in voltage and current levels, the integrated circuit of the chip contains regulation circuits which can be modelled at the first level by a variation of the load impedance Z_L of transponder. It has also effects on the resonance frequency. Moreover, at a given received power, losses will occur at the antenna level itself, mainly ohmic losses besides other losses from regulation circuits that cannot be approximated by a shunt resistance due to the presence of diodes that have no linear resistive behavior.

The first level electric circuit of the transponder is described in the following figure. It consists of two parts, the parallel RLC circuit, which is resonant due to the antenna and the tuning capacity of the chip, and the circuit containing active components of the RFID chip (Figure 2.23).

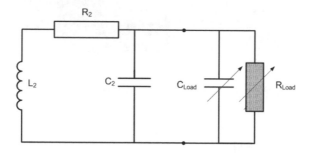

Figure 2.23. *Transponder electrical circuit*

The quality of a resonant circuit can be represented by the over-tension (or quality) factor Q which is equal to the ratio between the useful stored energy and the dissipated power in a period:

$$Q = \omega \cdot \frac{\text{Stored energy}}{\text{Dissipated power in a period}}$$

The magnetic loop circuit of the transponder antenna is equivalent to a parallel circuit $R_p L_2 C_2$. In fact, according to the series/parallel transformation rules of RLC networks, R_2 is considered as its parallel equivalent R_p by the rest of the circuit. The resistance R_p at resonance is:

$$R_p = R_2 \cdot (Q_0^2 + 1)$$

where Q_0 is the quality factor of the circuit, given by: $Q_0 = \frac{\omega \cdot L_2}{R_2}$.

Given E_{em}, the energy stored in the resonant circuit of the antenna with a voltage V_{chip}. E_{em} is written as:

$$E_{em} = \frac{1}{2} \cdot C_2 \cdot V_{chip}^2$$

The total dissipated power $P_{\text{total dissipated}}$ is divided into the power dissipated in the antenna P_0 and the power dissipated in the chip $P_{\text{active chip}}$ when it is activated (i.e. when the chip implements the following functions: supply and regulation, analog and digital data processing, reading and writing in non-volatile memory, load modulation, etc.).

$P_{\text{total dissipated}}$ is written as follows:
$P_0 + P_{\text{active chip}}$ and $P_{\text{total dissipated}} = \frac{\omega}{Q_{total}} \cdot \frac{1}{2} \cdot C \cdot V_{chip}^2$.

When the chip is not activated, i.e. below a voltage threshold that triggers the sequencing of a state machine or a minimal activity mode of the chip, we can consider $P_{\text{puce active}} = 0$.

In this case P_0 is determined by the relation:

$$P_0 = \frac{\omega \cdot C_2 \cdot V_{chip}^2}{2 \cdot Q_0}$$

$P_{\text{active chip}}$ is then given by the following relation:

$$P_{\text{active chip}} = \frac{\omega \cdot C \cdot V_{chip}^2}{2} \cdot \left(\frac{1}{Q_{total}} - \frac{1}{Q_0} \right)$$

We set:

$$Q_{\text{active chip}} = \frac{1}{Q_{total}} - \frac{1}{Q_0}$$

We obtain the relation $Q_{\text{active puce}} = \dfrac{\omega \cdot C_2 \cdot V_{\text{active chip}}^2}{2 \cdot P_{\text{active chip}}}$.

In general, in the specifications of RFID components, only the minimum and maximum electrical characteristics are provided. These characteristics are not sufficient to determine the active power. The RFID antenna design should have an electrical characterization step to measure the active power of the chip in its fundamental states (idle, active mode, reading and writing into memory, etc.).

2.3.4. *Design issue of RFID antennae in inductive coupling*

Because of the operating principle based on mutual inductance, when a transponder is in the operational zone of the reader and in its magnetic field, it appears as a load impedance. Its real and imaginary parts vary according to the coupling even when it is inactive as shown in Figure 2.24.

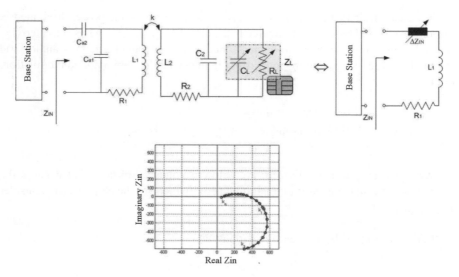

Figure 2.24. *Variation of the transponder load impedance according to the coupling*

This charge varies according to the received power, the execution of its operating system (or its state machine), writing steps in non-volatile memory, the load modulation, etc. The load of the chip cannot be represented by a single shunt resistor connected to a load capacity representing its electrical activity, since it contains active components which have no linear variation as a function of induced electrical quantities (current and voltage). The antenna design is therefore much more complicated.

Nevertheless several principal steps can be identified when designing antennae, taking into account issues of power transmission and data exchange:

– definition of the operational zone and initial geometry of reader/RFID transponder antennae: in this step, the minimum and maximum operating distances, the environmental constraints (metallic or non-metallic, mechanical, temperature, etc.), the configuration and geometry of base station antennae (antenna parameters R_1, L_1, C_1 or external dimensions) are part of the input data which should be taken into account. There are also electronic constraints of data processing (throughput and processing time) and operational electrical characteristics of RFID (optimal mode and the most unfavorable mode);

– determination of the coupling coefficient k according to different configurations of antennae in the operational zone: it is one of the essential steps in the antenna design since it determines the amount of remotely transmitted energy. The coupling coefficient will also affect the resonant frequencies of the circuits in mutual induction and thus will have an impact on the bandwidth of the system. This can be determined by experiment or by numerical simulation (Figure 2.25).

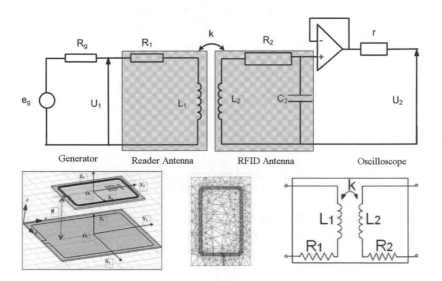

Figure 2.25. *Experimental measurement and simulation*
of the coupling coefficient

For experimental measurements, the amplitude of the generator should be selected, so that the generator does not supply the RFID chip and the generator remains in a state of high impedance. The follower circuit is used as "tampon circuit" so that the antennae are not disturbed. The coupling coefficient can be calculated by:

$$k = A_k \cdot \frac{U_2}{U_1} \cdot \sqrt{\frac{L_1}{L_2}}$$

A_k is a correction coefficient due to the influence of probe, cables and oscilloscope. Another possibility is to simulate the coupling coefficient in the operational zone by numerical simulation [COL 04]. Two antennae of the reader/RFID transponders form a quadrupole and the equivalent impedance matrix is written as:

$$Z = \begin{bmatrix} R_1 + j \cdot \omega \cdot L_1 & j \cdot \omega \cdot M \\ j \cdot \omega \cdot M & R_2 + j \cdot \omega \cdot L_2 \end{bmatrix}$$

where M is the mutual inductance. The matrix Z must be determined in different antenna configurations and positions. The numerical simulation uses the method of moments which is based on the numerical resolution of electromagnetism integral equations in which the unknown parameter is the surface current density. After building a triangular mesh to represent surfaces of arbitrary shapes, the common edges between the triangles are used to define basic functions (which are finite elements) defined by Rao, Wilton and Glisson (RWG) to determine the surface densities of current. Then by integration and resolution of potential vector (A) and scalar (V), we deduce a matrix algebraic system of admittance type [YV = I]. The impedance matrix Z is deduced by inverting the matrix Y.

The matrix Z is written as follows:

$$\begin{bmatrix} Z_{11} & Z_{12} \\ Z_{21} & Z_{22} \end{bmatrix}$$

and k is deduced from the matrix Z by the relation:

$$k = \frac{Im(Z_{12})}{\sqrt{Im(Z_{11}) * Im(Z_{22})}}$$

The expected results for this step are the determination of an acceptable variation range for the coupling coefficient and then determining the voltage received at the terminals of the RFID chip antenna as a function of the electrical quantity (current or voltage) of the reader antenna. The voltage variations at the terminals of the chip can be critical in case of over-voltage or in the case where signal to noise ratio is too low for ensuring reliable communication between the RFID chip and the reader. In all cases, these voltage variations should be compatible with the electrical specifications prescribed by the specification of the component.

For good reliability of energy transmission, we determine the transfer functions of electrical quantities (usually a voltage across the reader antenna) for the output electrical quantity (usually the voltage across the chip) and then trace the variations of these transfer functions to privilege the low variation zones.

– Contactless functional characterization and measurement of power consumed during a transaction, including command for writing and reading with a cryptographic implementation. It is a fundamental step to determine exhaustively the electrical behavior of an RFID chip. The main difficulty with this characterization is that the electrical behavior of the RFID chip is not linear. It establishes an electrical model which is representative of the chip in a transaction within optimal conditions and in the worst case (in connection with the distribution of magnetic field lines, and the coupling in the operational zone).

An example of an electric model is described in Figure 2.26.

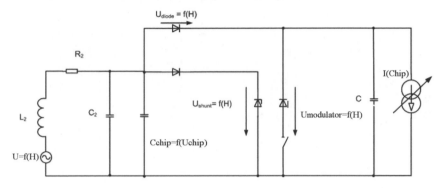

Figure 2.26. *Circuit equivalent to an RFID chip*

The expected result in this step includes determination of the total quality coefficient of the RFID chip and the charge effect induced into the reader antenna.

– Determination of optimal operational zone which guarantees supply (current-voltage) of the chip for the chosen range of variation of coupling coefficient with electrical parameters set for the reader antenna emitting a field between the values set by the ISO standard. Based on the calculations defined in the previous steps, for a given voltage variation range corresponding to a defined operating mode, we can draw networks of curves (R2, L2) which meet chosen criteria. This step also ensures proper data transmission with the corresponding throughput, by taking into account the total quality factor of the active chip in terms of electric and communication.

2.3.5. *Far field coupling*

A UHF RFID system is governed by an equation called "of telegraphists" following the theory of transmission lines. It shows that as the wavelength of used frequencies is of the same order of magnitude as the size of antennae, incident and reflected electric waves coexist. We use the formalism of power waves which are measurable by scattering parameters S [GHI 08].

2.3.5.1. *Reflection coefficient*

In a voltage-current field, a dipole is characterized by its impedance Z (the ratio between voltage and current). Its equivalent in the formalism of parameters S is the reflection coefficient Γ which is the ratio between the reflected wave and the incident wave and is representative of the fraction of energy reflected by the dipole Z (Figure 2.27):

$$|\Gamma|^2 = \frac{P_r}{P_i}$$

where P_i and P_r are respectively the incident and reflected electrical power.

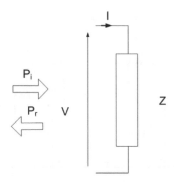

Figure 2.27. *Reflection on a dipole Z*

We define the reflection coefficient of an antenna by the following relation:

$$\Gamma = \frac{Z_a - Z_0}{Z_a + Z_0}$$

Z_0 represents characteristic impedance (typically 50Ω) for which there is no reflected wave and Z_a is the antenna impedance.

The reflection coefficient is quantified in amplitude, and is in phases the reflected energy by the dipole. It also uses a different formalism to measure the reflected energy, but without giving any information on the phase of the reflected signal: the stationary wave rate or TOS, given by:

$$TOS = \frac{1 + \Gamma}{1 - \Gamma}$$

In general, we try to maximize the power transmitted to the load. We define the coefficient of reflection in power Γ^* by the relation:

$$\Gamma^* = \frac{Z_l - Z_a^*}{Z_l + Z_a}$$

Z_l represents the load impedance of the chip connected to the antenna impedance Z_a.

Γ^* represents the ratio between the power reflected by a load and the maximum total power that can be transmitted to it. This is the classic case of impedance adjustment where when the load is complex conjugate with the impedance of the antenna ($Z_l = Z_a^*$), where the maximum power transmission (i.e. $\Gamma^* = 0$) corresponds to half the total power received by the antenna. The other half is radiated by the radiation resistance R_r.

2.3.5.2. Transmission coefficient

We define the power transmission coefficient T by:

$$T = 1 - |\Gamma^*|^2$$

It corresponds to the ratio of power transferred to a load and the maximum total power which can be transferred to it. T can be written as follows:

$$T = \frac{4 \cdot R_a \cdot Rl}{\cdot |Z_l + Z_a|^2}$$

2.3.5.3. Quality coefficient and bandwidth

As we have seen, the quality coefficient Q of a resonant circuit is defined by the following equation:

$$Q = 2 \cdot \pi \cdot \frac{\text{stored energy per cycle}}{\text{radiated and dissipated energy per cycle}}$$

The radiated and dissipated energy correspond to the lost energy due to conduction and dielectric losses.

The coefficient Q indicates whether an antenna is capable of storing energy and determines its frequency selectivity by the relation:

$$Q = \frac{f_0}{\Delta f}$$

where f_0 is the resonance frequency and Δf the bandwidth, that is defined as the difference between the frequencies for which the resistance is equal to the reactance.

However, if the TOS is considered, to take into account the possible mismatch of the antenna, we obtain the following equation:

$$Q = \frac{f_0 \cdot \sqrt{TOS}}{\Delta f \cdot (TOS - 1)}$$

2.3.5.4. *Directivity and gain*

An antenna (electric doublet) transmits or receives most of its energy perpendicular to the wire and a small amount of energy is transferred along the length of the wire. The directivity is a measure of the capacity of an antenna to concentrate the radiated power. Such a directivity characteristic is one of the most important qualities of an antenna. It can direct the signal transmission or reception in a given axis, and exclude transmission or reception in other directions.

The directivity in a given direction, $D(\theta, \varphi)$, is the ratio between the radiated power density in this direction and the total power density which would be radiated by a lossless isotropic antenna (reference antenna radiating uniformly in all directions of space):

$$D(\theta, \varphi) = 4 \cdot \pi \cdot \frac{\text{radiated power density in the direction}(\theta, \varphi)}{\text{total power density}}$$

Note that an isotropic antenna, by definition has a directivity equal to unity, in all directions.

The second fundamental quality of an antenna is its gain. The intrinsic gain of an antenna in a given direction, $G(\theta, \varphi)$, is given by the ratio between the power density caused by the actual considered antenna in this direction where the density is maximum, and the power density generated under the same conditions of total radiated power, by a lossless isotropic hypothetical antenna:

$$G(\theta, \varphi) = 4 \cdot \pi \cdot \frac{\text{radiated power density in the direction}(\theta, \varphi)}{\text{total provided power}}$$

In the formulation of the gain, we take into account only the antenna losses but not the mismatch loss (or insertion loss).

Note also that the notion of antenna gain does not denote an "active" character, but rather a comparison of performance.

2.3.5.5. *Radiation resistance*

The radiated power $P = 40 \cdot (\frac{\pi \cdot I \cdot dl}{\lambda})^2 = 40 \cdot (\frac{\pi \cdot dl}{\lambda})^2 \cdot I^2$. This expression is of the form $P = \frac{1}{2} \cdot R \cdot I^2$, which expresses a power dissipated through a resistance R carrying a current of amplitude I, so $R_r = 80 \cdot (\frac{\pi \cdot l}{\lambda})^2$. R_r is the radiation resistance. If we neglect the ohmic losses in the conductors, R_r represents the resistance of equivalent load for the source in the case of emission or the internal resistance of the equivalent generator which constitutes the reception antenna.

2.3.5.6. *Radiation efficiency*

The total radiation efficiency, η_t, of an antenna is a multiplicative factor that is used to take into account losses at the antenna's input, as well as in its structure. Indeed, the losses can be caused by reflections from a bad adjustment of the antenna to the associated circuit, as well as by the properties of dielectric materials and conductors that constitute it.

η_t is given by:

$$\eta_t = \eta_r \cdot \eta_{cd}$$

where η_r is the efficiency coefficient due to the mismatch and η_{cd} is the coefficient due to conduction and dielectric losses [GHI 08].

The power radiated by an antenna, P_{rad}, is linked to the power accepted by the antenna, P_{in}, through η_{cd} by the relation:

$$P_{rad} = \eta_{cd} \cdot P_{in}$$

The relations between the gain of an antenna and its directivity is given as:

$$G(\theta, \varphi) = \eta_{cd} \cdot D(\theta, \varphi)$$

We can also define the quality coefficient of radiation Q_{rad} as a function of the quality coefficient Q of an antenna as:

$$Q = \eta_{cd} \cdot Q_{rad}$$

If we increase the losses of an antenna, the quality coefficient of the antenna reduces and the bandwidth increases.

The efficiency η_{cd} depends on the parameters of the antenna circuits. It is defined as the ratio between the power dissipated by the radiation resistance and the total dissipated power:

$$\eta_{cd} = \frac{R_r}{R_r + R_l}$$

2.3.5.7. *Radiation diagram*

The radiation diagram represents the distribution of the electromagnetic field radiated into space in a given direction. Therefore, it provides information on the capacity of an antenna to radiate into space. From the radiation diagram, it is possible to define several radiation parameters of an antenna such as aperture, power level and direction of secondary lobes. There are a multitude of ways to represent the radiation of an antenna: field diagram, in power, gain, directivity, in polar or Cartesian, in linear or decibels, 2D or 3D, etc.

2.3.5.8. *Polarization*

One of the particularities of far fields is that the ratio between the amplitude of the electric fields and the amplitude of the magnetic field is constant. We need to simply measure one to evaluate the other. Polarization corresponds to the direction and the amplitude of the electric field which forms the electromagnetic wave. It shows a geometry drawn by the tip of the electric vector $\vec{E}$ in time (Figure 2.28). It is important to know the polarization which is one of the fundamental parameters of antennae to determine how the antenna transmits or receives the electric field in an authorized way.

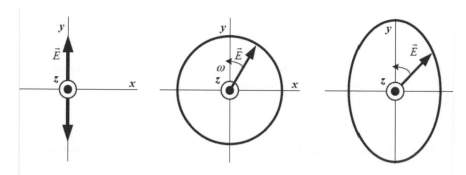

Figure 2.28. *Polarization of electromagnetic wave*

There are three types of polarizations:

– linear polarization (Figure 2.28): an electromagnetic wave is linearly polarized if at each moment its electric field is oriented in the same direction. It corresponds to an ellipticity ratio $AR = \infty$;

– circular polarization (Figure 2.28): an electromagnetic wave is circularly polarized if its electric field describes a circle in time. It corresponds to an ellipticity ratio $AR = 1$. However, in general, the circular polarization corresponds to $1 < AR < 2$;

– elliptic polarization (Figure 2.28): an electromagnetic wave is elliptically polarized if its electric field describes an ellipse in time. It corresponds to an ellipticity ratio $1 < AR < \infty$.

In the case of linear polarization, it is called vertical polarization when the electric field is vertically oriented relative to the horizon, and horizontal polarization when the electric field is oriented in a plane parallel to the horizon.

In general, the polarization of an antenna cannot be the same as the polarization of an incident wave.

Given:

$$\vec{E}_i = \hat{p}_i \cdot \left| \vec{E}_i \right|$$

and:

$$\vec{E}_a = \hat{p}_a \cdot \left| \vec{E}_a \right|$$

$\hat{p}_i$ and $\hat{p}_a$ respectively represent the unit complex vectors of electric fields of the incident wave and the antenna.

The reception quality of an electromagnetic wave by an antenna depends on the polarization difference of the receiving antennae compared to one of the transmitting antennae and its adaptation to the orientation of the electric field of the received wave.

We define Polarization Loss Factor (PLF) by the following formula: $PLF = |\hat{p}_i \cdot \hat{p}_a|^2$ where PLF is directly proportional to $cos^2\alpha$ constituted by angular difference of polarization between the incident signal and the antenna (Figure 2.29).

Label Orientation	Reader Antenna Polarisation		
	circular	vertical	horizontal
vertical	3dB	0dB	∞ dB
horizontal	3dB	∞ dB	0dB
slant	3dB	3dB	3dB
parallel to antennabeam	∞ dB	∞ dB	∞ dB

Figure 2.29. *Polarization mismatch*

2.3.5.9. *Transmission in free space (Friis equation)*

Transmission in free space is governed by the Friis equation and is also called the link budget equation, where the equation calculates the power available at the load level during reception, according to the power supplied to the antenna in transmission. It is given by the following equation:

$$\frac{P_r}{P_t} = G_t \cdot G_r \cdot \eta_r \left(\frac{\lambda}{4 \cdot \pi \cdot r} \right)^2$$

with:

P_r: transmitted power delivered at the transmission antenna,

P_t: received power at the reception antenna,

R: distance between the transmission antenna and the reception antenna,

λ: wavelength,

G_t:linear gain of the transmission antenna,

G_r: linear gain of the reception antenna.

The antenna gains are measured relative to an isotropic antenna. The measurement does not take into account adjustment losses or performances during transmission and reception.

The term $(\frac{\lambda}{4 \cdot \pi \cdot r})^2$ is called Path Loss (PL). PL est usually given in decibels:

$$PL(indB) = 32.5 + 20 \cdot log_{10} r_{km} + 20 \cdot log_{10} f_{MHz}$$

In addition, there are additional attenuations due to diffraction and absorption.

If we take into account adjustment losses and performances of antennae in transmission and reception, a more complete formulation would be:

$$\frac{P_r}{P_t} = \eta_{cdt} \cdot \eta_{cdr} \cdot (1 - |\Gamma_t|^2) \cdot (1 - |\Gamma_r|^2) \cdot \left(\frac{\lambda}{4 \cdot \pi \cdot R}\right)^2 \cdot D_t(\theta_t, \varphi_t) \cdot D_r(\theta_r, \varphi_r) \cdot |\hat{p}_t \cdot \hat{p}_r|^2$$

2.3.5.10. *Propagation in free space*

When an antenna generates radiated power P_e, we want to determine the received power P_r. We assume that the antennae are properly oriented, i.e. the maxima of the directivity diagrams are aligned.

If the transmission antenna were isotropic, it would radiate throughout the space with a power density:

$$p_{iso} = \frac{P_e}{4 \cdot \pi r^2}$$

If the antenna has a gain G_e, the radiated power density in the direction of the maximum of the directivity diagram is: $p = p_{iso} \cdot G_e$ where:

$$p = \frac{P_e \cdot G_e}{4 \cdot \pi r^2}$$

The power density created in a given direction is the product of the gain in this direction and the power. This product $P_e \cdot G_e$ is called EIRP (Equivalent Isotropically Radiated Power) which is the reference value used in standard definitions (Figure 2.30). It is the power that the transmitter should have, if its antennae were isotropic, so as to get the same result at the same distance. Another reference can be used in standards dedicated to UHF radio communications such as ERP (Effective Radiated Power) which is the product of the power radiated and the gain relative to an antenna of dimension of half wavelength. The relation between ERP and EIRP is given by:

$$EIRP = ERP \cdot 1,64.$$

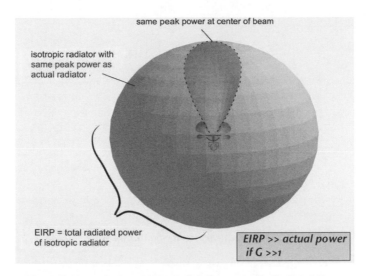

Figure 2.30. *Definition of Equivalent Isotropically Radiated Power*

2.3.5.11. *Equivalent electrical circuit of an antenna*

As we have seen previously, the electrical parameters of antennae (electric or magnetic dipole) have a reactance (inductive or capacitive) that predominates over the resistive part. They present radiation characteristics which are inefficient and difficult to adapt. Nevertheless, they are suitable for near fields but are unsuitable for far fields where the notions of efficiency in transmission and reception is important. Due to the small size of wavelengths relative to the size of circuits, the dimensions of resonant antennae are selected in the order of half or quarter wavelength.

The equivalent circuit of a transmission antenna is shown in the following Figure 2.31:

The antenna is characterized by $Z_a = R_a + j \cdot X_a$.

R_a is the dissipated energy and is constituted of the radiation resistance R_r, which represents the radiated energy, while the loss resistance R_l represents conduction losses, dielectric losses and losses due to the skin effect. At transmission, a generator connected to the antenna has an output impedance Z_g which consists of a resistive part R_g and a reactance X_g.

In case of reception, the equivalent circuit of the antenna is shown in the following Figure 2.32:

The antenna is connected to a load with an impedance Z_{chip} with $Z_{chip} = R_{chip} + j \cdot X_{chip}$. An application of the Friis equation is that the voltage at the terminals of

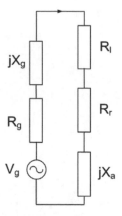

Figure 2.31. *Equivalent circuit of a transmission antenna*

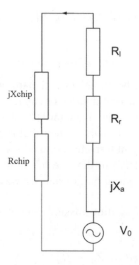

Figure 2.32. *Equivalent circuit of an antenna in reception*

the antenna of the RFID chip V_0 is written as a function of the antenna gain and the incident power density:

$$\frac{|V_0|^2}{8 \cdot R_a} = S_i \cdot \frac{\lambda^2}{4 \cdot \pi} \cdot G_r$$

where G_r is the gain of the reception antenna. The incident power density is the result of the expression of Poynting and is given by:

$$S_i = \frac{G_t \cdot P_t}{4 \cdot \pi \cdot r^2}$$

2.3.5.12. *Radar Cross-Section*

A reception antenna will receive power that is proportional to the emitted power. The proportionality coefficient is homogeneous in a surface that is defined as the equivalent surface of an antenna.

Note that this surface is, in most cases, fictious since the reception antenna can also be a simple conductor of negligible real surface and is not linked to the geometric dimensions of the structure.

In addition, following the principle of reciprocity, a conductor can be considered as a transmission antenna as well as a reception antenna, since it is supplied by a source or loaded by a load resistance. This basic principle shows that a transmission antenna should have the same characteristics during reception and give the same results when calculating the ratio of the received voltage to the inductor current. We then obtain the relation between two features gain and equivalent surface:

$$\frac{G}{A_e} = \frac{4 \cdot \pi}{\lambda^2}$$

Regarding UHF RFID systems, as we have seen, the principle of communication, when the RFID chip exchanges data with the reader is based on the commutation of impedances representing the state changes. Given that these signals are part of the far field zone, the reader detects the reflected waves coming from the impedance change. We want to determine a fundamental characteristic of these waves which is Radar Cross-Section (RCS), where RCS is defined as the equivalent surface, previously introduced and which, by intercepting an amount of power P_t and radiating isotropically, produces the same density of reflected power P_r by the target at the receiver.

Recall that in the case of an isotropic transmitter, the emitted energy is distributed over the surface of a sphere of radius R. Given S_i, the power density of the incident wave and S_r, the power density of the reflected wave, the radar section is given by:

$$\sigma = R \overset{lim}{\to} \infty \left[4 \cdot \pi \cdot R^2 \cdot \frac{S_i}{S_r} \right]$$

As dipole antennae generally have weaker sections and a length lower than the half-wavelength, it is shown that:

$$\sigma = |1 - \Gamma^*|^2 \cdot \frac{\lambda^2}{4 \cdot \pi} \cdot G^2$$

Given the factor K which is defined by the relation: $K = |1 - \Gamma^*|^2$, K can also be expressed in terms of the impedance of the antenna and its load:

$$K = \frac{4 \cdot R_a^2}{|Z_a + Z_{chip}|^2}$$

The equation of σ reveals the influence due of the adjustment between the load and the antenna on the RCS and shows the need to determine parameters of the chip, i.e. its electrical characterization, since the measurement of RCS does not help to find the impedance of an antenna. In the case of the half-wavelength dipole, the antenna gain is known and the antenna impedance is purely real. So if the antenna is adapted to radiate at 868 MHz, its RCS can easily be calculated irrespective of the load at terminals.

2.3.5.13. *Radar cross-section modulation*

The equation for the radar cross-section σ shows that by modulating the load of the chip, the equivalent surface of the transponder will vary and cause an amplitude modulation on the reflected signal to the reader.

A load modulation between the impedances Z_{L1} and Z_{L2} corresponds respectively to the reflection coefficients σ_1 and σ_2. As the reflected field is proportional to the current in the antenna, the difference of reflected power corresponds to the power radiated by the antenna where the radiated power is due to the difference between complex currents I_1 and I_2 corresponding to the two impedance states Z_{L1} and Z_{L2}. We obtain, according to the equivalent circuit of the antenna:

$$I_{1,2} = \frac{V_0}{Z_a + Z_{L1,L2}} = \frac{V_0}{2 \cdot R_a} \cdot (1 - \Gamma_{1,2})$$

The modulation of the radar cross-section is given by:

$$\Delta\sigma = \frac{\lambda^2 \cdot G}{4 \cdot \pi} \cdot |\Gamma_1^* - \Gamma_2^*|^2$$

that we want to optimize.

The quality of the modulation depends on the possibility to differentiate two binary states during demodulation. For an amplitude modulation, the ideal case would be to have a maximum amplitude difference between the high and low states. It is achieved, for example, when one moves from the perfect adjustment of the antenna to the case where the antenna is short-circuited. However, it implies that when there is total reflection of incident wave, the energy recovery by the chip becomes impossible. Therefore the index of the ASK modulation is generally chosen to achieve a balance between energy recovery and data transmission quality.

In conclusion, the Friis equation and the preceding one show that the performance of a UHF RFID systems, in terms of reading distance and modulation of the signal carrying the information, depends on the following parameters:

– adjustment quality between chip and antenna;

– antenna efficiency by minimizing conduction and material losses;

– gain;

– directivity.

2.4. Bibliography

[CAU 05] CAUCHETEUX D., BEIGNÉ E., RENAUDIN M., , CROCHON E., "Towards Asynchronous and High Data Rates Contactless Systems", *PRIME'05*, Lausanne, Switzerland, July 2005.

[COL 04] COLOMBO M., BARBU S., ELRHARBI S., "EM Simulation of 13.56 MHz Inductive-Coupled RFID Antennae by MoM: Novel Impedance-Matrix Approach", *MS'04*, Marseille, France, July 2004.

[ENG 03] ENGELS D., The Use of the Electronic Product Code, Research report, Massachusetts Institute of Tech, May 2003.

[GHI 08] GHIOTTO A., VUONG T., TEDJINI S., WU K., "Design of Passive Ultra-High Frequency Radio-Frequency Identification Tag", *URSI'08*, Chicago, USA, 7-16, August 2008.

[PEA 07] PEARSON J., MOISE T., The Advantages of FRAM-Based Smart ICs for Next Generation Government Electronic IDs, Research report, Texas Instruments, Inc., September 2007.

Chapter 3

RFID Communication Modes

3.1. Communication modes

3.1.1. *Waveforms and usual communication codes of RFID systems*

There are two main communication protocols between a RFID tag and a RFID reader:

– in the TTO (*Tag Talk Only*) protocol, only the RFID tag transmits data. There is no uplink. The RFID tag transmits its data regularly when it is provided;

– in the RTF (*Reader Talk First*) protocol, the reader is the master of the communications. It is used in most RFID technologies, including the EPCglobal Class 1 Generation 2 standard. Typically, when an RFID tag enters into the magnetic field of a reader, it waits for a request from the reader before transmitting its identity code.

As seen previously, there are two types of processes for communications between the reader and the RFID tag: processes using continuous energy transfer and processes transferring energy in a sequential manner.

In the case of a process of continuous energy transfer and data transmission in *Half Duplex* (HDX) mode, the communication protocol between a reader and an RFID tag is composed of three phases:

– recovery phase of the RFID tag: the reader sends an electromagnetic wave to the RFID tag to provide it with needed energy for its operation. Then the tag stays at a waiting state until it receives an instruction from the reader;

Chapter written by Simon ELRHARBI and Stefan BARBU.

– instruction phase: the reader sends an instruction (typically a request) to the RFID tag;

– reading phase: the RFID tag sends its reply corresponding to the request of the reader.

Most of the RFID systems transform their signals using frequency translation and transcoding. The choice of signal modulation and data encoding is a key element for the reliability of RFID data transmission. Indeed, the modulation type and the mode of communication in both transmission directions determine not only the system bandwidth, the electric power transmission, the data integrity, but also the complexity of electronic emission architectures and RFID signal receptions.

Due to the low hardware and software resources embedded in RFID tags and regulatory restrictions on the emission of radio power for readers, the choice of signal modulation and data coding is limited. Nevertheless, there still remain possibilities on the return path for signal encoding on the RFID tag side since RFID does not act in an active way (in terms of radio signal emission) but acts in a passive way. By consequence, RFID is not in the framework of radio emission regulations.

3.1.2. *Data coding*

Different coding types used in RFID systems are given in Figure 3.1.

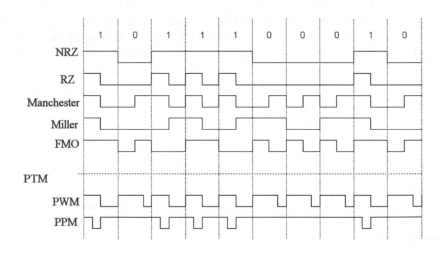

Figure 3.1. *Example of coding schemes used in RFID systems*

Two main coding types of baseband signals are used:

1) the coding by level in which a binary value corresponds to a voltage level of the signal. For example, one of the simplest coding methods is the NRZ (for *Non-Return-to-Zero*) code that associates the high logic level with the binary value 1 and the low logic level to the binary value 0. To avoid a series of bits "1" or "0", which can result in signal desynchronisation and difficulty to reconstruct the time base, the signal drops back to 0 V after each binary value. This is the RZ (*Return-to-Zero*) coding. A positive pulse corresponds to a binary value 1 and its duration is equal to half the symbol time. In the other case, the signal stays at the low logic level. Moreover, the encoding (NRZ or RZ) of data is independent of the coding of the previous data,

2) the coding by signal transition in which a binary value corresponds to a variation between two voltage levels of the signal; information is contained in the transitions of the signals and the data encoding is based on the encoding of the previous data. In this case, the clock synchronization is easier as compared to synchronization in the coding level.

- The Manchester coding is an example of the coding by transition. It designates a binary value by the sign of the edge at the half-period instant of each bit. Thus, a rising edge in the middle of the symbol time corresponds to the binary value 0 and a falling edge corresponds to the binary value 1. The bit rate is equal to the communication bandwidth. The transition in the middle of the symbol time is vital for signal synchronization at the reception and in particular for anti-collision detection processes when multiple cards using this coding type are in the operating area of the RFID reader. However, the Manchester code has a higher bandwidth than other coding methods. In addition, during the Manchester coding it is ensured that the encoding of data is independent of the coding of the previous data.

- Another type of coding, known as "Miller code" uses a transition at half the symbol time to encode the binary value 1. If a sequence of 0 bits happens, a transition at the beginning of the symbol time is added. This ensures a change in logic level at least after a period that is equal to two symbol times. A variant of this coding, called "modified Miller code" uses the same principle as above but with each transition replaced by a negative pulse. The bandwidth required to transmit this code is much more important than in the previous case. During Miller encoding, the encoding of the previous data is considered.

- FM0 (or *Bi-Phase Space*) is similar to the Miller coding, as it reverses the phase at the beginning of each period of symbol time while for the bit 0, it adds a phase inversion in the middle of the symbol time. This encoding, as compared to the Miller coding, allows better synchronization during reception. Identical to the Miller coding, the FM0 encoding of data considers encoding of the previous data.

Other coding schemes transmit information by Pulse Time Modulation (PTM). In this category, we use Pulse Width Modulation (PWM), Pulse Position Modulation (PPM) and Pulse Interval (PIE) coding methods:

– *Pulse Width Modulation PWM*. The pulses are regularly spaced at a constant amplitude and at a duration proportional to the amplitude of the signal. The PWM code links the binary value to the length of the positive pulse. At the end of the symbol time, the level is always returned to the low state before moving on to the high state for each new encoded bit.

– *Pulse Position Modulation PPM*. Information is coded depending on the position of the pulse. The PPM code uses a negative pulse to encode the bit logic 1. However, unlike the case of modified Miller code, a series of logical bits 0 is always encoded by a constant level at the high state. By extension, the PPM codes with of order n encode logic words of n bits. The position of the pulse in the time period allows us determine the codeword. However any word will be coded by the position of a negative pulse: it is not possible to find a constant level throughout the symbol time. This type of code, as compared to the Manchester code, has a relatively narrow bandwidth and is easy to implement. However it suffers from low data rates.

– *Pulse Interval coding PIE*. It is a variant of the PPM modulation. The reader creates two pulses whose falling edges define the interval. This interval varies as a function of binary values 0 and 1.

Coding techniques for RFID information should consider the following constraints:

– the code should maintain power transmission as long as possible;

– the code should not consume too much of bandwidth;

– the code should help in collision detection if there are several RFID tags in the operating area of the reader.

The PPM, PIE and PWM codes, due to their relative signal stability, satisfy the first two constraints. However, the implementation of the Manchester code, as we shall see further, allows easy detection of bit collisions. Therefore this coding type is often used for the return path, when RFID tags send data to the reader.

Codes with the lowest bandwidth are NRZ and Miller codes. They have a bandwidth that is about half the bandwidth of data bit rate. They are followed by the Manchester, FM0 and RZ codes which have a bandwidth equal to that of the communication throughput.

The coding choice for the binary representation should be made according to the need for remotely supplying the RFID while keeping as long as possible the presence of the carrier signal (which enables the NRZ or Miller coding). Regarding the dialog between the RFID tag and the reader, it is important to detect the response contained in the return signal. A coding that has a transition during the bit time (Manchester) simplifies the task.

3.1.3. *Modulation*

After choosing data representation by suitable coding, the next step is to determine the choice of waveforms of signals that will be exchanged between the RFID tags and the readers. Conventionally, communications consist of a carrier signal which can be modulated to transport the binary information in three possible ways:

– Amplitude Shift Keying (ASK) where the amplitude of the carrier signal is modulated between several levels;

– Frequency Shift Keying (FSK) where the frequency of the carrier signal is modulated between several frequencies;

– Phase Shift Keying (PSK) where the phase of the carrier signal is modulated between multiple values.

Each modulation type has its own power distribution. The ASK modulation requires the carrier to be transmitted at high power, which can be detrimental for embedded systems like RFID technologies. The advantage of the ASK modulation is the simplicity of the demodulation architecture. The FSK modulation can be seen as a special case of PSK modulation. It has an advantage in energy distribution since it is distributed almost (98%) throughout the frequency band but has relatively complex reception architecture.

However, a significant difference in power distribution between readers and chips causes a fundamental constraint in RFID technologies. The bandwidths of codes should be limited, so that the bandwidth are compatible with the specifications of the reader. Indeed, when the quality factor of the antenna in the transmitter circuit is higher, the transmission power will be optimal but the bandwidth will be reduced, thus limiting the communication throughput.

In addition, the return signal from the RFID tag to the reader can be merged with the forward signal, so that it may be necessary to modulate the return signal on another frequency or create a sub-carrier by impedance modulation with a frequency that is different from the carrier frequency.

Figure 3.2 shows a summary of the modulation and coding methods used in actual RFID standards.

The hardware architecture of the RFID tag is a balance between the complexity of its architecture (related to its security and features), its size and its cost that must be low because of mass deployment in different physical and electrical environments with many constraints. Moreover, the radio hardware interfaces of the tag and the reader must be robust and simple.

Characteristics of the transmission link				
Standard	Type	Carrier	Modulation	Coding
18000-2	A (FDX)	125 kHz	ASK (100%)	PIE
	B (HDX)	134.2 kHz		
18000-3	Mode 1	13.56 MHz	ASK (10%) and ASK (100%)	PPM
	Mode 2		PSK - PJM (Phase Jitter Modulation)	FSK - MFM (Modified Frequency Modulation)
18000-4	Mode 1	2.40 – 2.48 GHz	ASK (90% - 100%)	Manchester
	Mode 2		FSK - GMSK (Gaussian Minimum Shift Keying)	
18000-6	A	860 MHz – 960 MHz	ASK (27% - 100%)	PIE
	B		ASK (18% or 100%)	Manchester
EPCglobal	Class-1 Generation-2	860 MHz – 960 MHz	ASK (90%): – DSB (Double-SideBand) – SSB (Single-SideBand) – PR (Phase-Reversed)	PIE
18000-7		433.92 MHz	FSK	Manchester
15693-2		13.56 MHz	ASK (10%) and ASK (100%)	PPM
14443	A	13.56 MHz	ASK (100%)	Miller Modified
	B		ASK (10%)	NRZ

Characteristics of the reception link				
Standard	Type	Sub Carrier	Modulation	Coding
18000-2	A (FDX)	No	FSK	Manchester
	B (HDX)	134.2/123.7 ± 4 kHz		NRZ
18000-3	Mode 1	432.75 kHz or (2 sub-carriers in phase relationship)		Manchester
	Mode 2	423.75 kHz and 484.28 kHz		
18000-4	Mode 1	No	ASK	FM0
	Mode 2	Notification: 153.6 kHz Communication 384 kHz	PSK or OOK (ASK)	
18000-6	A	No	ASK two states	FM0
	B			
EPCglobal	Class-1 Generation-2	40 kHz to 640 kHz	ASK and/or PSK	FM0 or Miller sub carrier
18000-7		No	FSK	Manchester
15693-2		423.75 kHz or two sub carriers at 423.75 and 484.28 kHz		Manchester
14443	A	847 kHz	ASK	Manchester
	B		BPSK	NRZ

Figure 3.2. *Summary of the signal modulation and coding methods used in RFID standards*

3.1.4. *Integrity of transmissions in RFID systems*

During transmissions of energy and data in the radio channel, various electromagnetic sources can cause transmission errors. In order to ensure the integrity of transmissions, it is necessary to detect the presence of errors in the received message. If this happens, the protocol should be able to transmit the message again unless the receiver can correct the message itself.

There are essentially three types of error detection modes:

– Vertical Redundancy Check (VRC) that is also called the parity check;

– Longitudinal Redundancy Check (LRC);

– Cyclic Redundancy Check (CRC).

3.1.4.1. *VRC process*

The VRC process or parity check is a very simple process. The principle is to add an extra bit (called the parity bit) to the number of data bits to form a byte. The parity bit is equal to 1 if the number of data bits is odd and 0 otherwise. In general, this type of parity check uses XOR (exclusive OR) logic gates. At the reception, the parity of the message is calculated: if it matches the chosen format, the receiver considers that there has been no transmission errors. Otherwise, an error is detected and the message will be transmitted again. So a relatively low redundancy indication is added to messages. However, the simplicity of this method only allows a very low detection rate. Indeed, if there is an even number of transmission errors, they are not detected. An odd number of errors will be detected during transmission of the byte [CAU 05]. Parity checking is not used in RFID standards but is still valid for proprietary systems having RFID chips with low computational resources.

3.1.4.2. *LRC process*

The LRC (Longitudinal Redundancy Check) process, also called cross parity check, not only controls data integrity of a character, but also checks integrity of parity bits of a block of characters. It is based on a recursive method. The byte 1 (to transmit) and the byte 2 are bitwise XORed. The obtained result is processed by the same operation with the byte 3, and so on with all the bytes of the message. Once the LRC byte is obtained, it is transmitted with data. The receiver of the message performs this operation again, but this time takes into account an extra byte: the LRC byte. If the result is 00, then no error is detected.

The simplicity of this algorithm ensures that the result is computed quickly and easily. However, the LRC process is not very reliable as multiple errors can be cancelled by themselves. That is why this process is used for quick checks on small data sizes. The cross parity check is not used in RFID standards, though it still remains available for designers of proprietary RFID systems.

3.1.4.3. *CRC process*

The CRC process is used to handle large data sizes with high reliability of error detection. It is based on a mathematical algorithm to detect and, if necessary, repair the errors in data in frames. The principle of the CRC process is to manipulate binary sequences as binary polynomials, i.e. polynomials whose coefficients correspond to the binary sequence. The least significant bit (LSB) of the sequence represents the degree 0 of the polynomial ($X^0 = 1$) and the nth bit represents the degree $n - 1$ of the polynomial. Then the division of the data (usually extended by the bits 0) is performed by the polynomial. The remainder of this division is then transmitted with data to the communication receiver. The latter performs the same step and obtains 0 as the rest state if there are no transmission errors. This explains the presence of the rest state in the division during transmission. For this error detection mechanism, the organisation of standard and recommendations in the Telecommunications field (CCITT), replaced now by ITU-T (International Telecommunication Union) recommends some polynomials for different frame types. Figure 3.3 presents a summary of the polynomials used in the main RFID standards.

Standard	CRC Length (bits)	Polynome
18000-2	CRC-16	$X^{16}+X^{12}+X^5+X^0$
18000-3	CRC-32	$X^{32}+X^{26}+X^{23}+X^{22}+X^{16}+X^{12}+X^{11}+X^{10}+X^8+X^7+X^5+X^4+X^2+X^{26}+X^1+X^0$
18000-4	CRC-22	$X^{22}+X^{17}+X^{13}+X^9+X^4+X^0$
	CRC-15	$X^{15}+X^{10}+X^9+X^6+X+X^0$
	CRC-44	$X^{44}+X^{30}+X^{29}+X^{15}+X+X^0$
18000-6	CRC-5	$X^5+X^3+X^0$
	CRC-16	$X^{16}+X^{12}+X^5+X^0$
EPCglobal	CRC-5	$X^5+X^3+X^0$
	CRC-16	$X^{16}+X^{12}+X^5+X^0$
18000-7	CRC-16	$X^{16}+X^{12}+X^5+X^0$
14443	CRC_A	$X^{16}+X^{12}+X^5+X^0$
	CRC_B	

Figure 3.3. *Synthesis of CCITT polynomial used for CRC calculation*

For example, for a 16-bit CRC calculation, we use the CRC-CCITT16 polynomial (Figure 3.4):

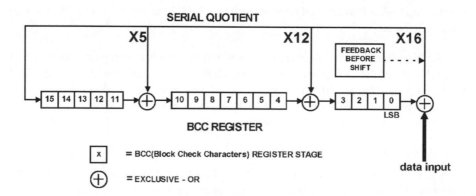

CRC-CCITT GENERATING POLYNOMIAL = X16 + X12 + X5 + X0

Figure 3.4. *16-bit CRC process*

$$P(X) = X^{16} + X^{12} + X^5 + X^0$$

with an initial value of 0x0000.

The 16-bit CRC shift register is initialized to zero at the beginning of the process. The LSB (Least Significant Bit) input data is XORed with the MSB (Most Significant Bit) bit of the CRC register, and is then shifted in the LSB register. At the end of the process, the CRC register incorporates the result of the CRC-16 code.

From a practical point of view, the CRC of a message is obtained by updating the CRC, where the CRC handles the message byte by byte in two ways:

– method with table generation: this method consists in storing in memory a table containing the CRC (2 bytes) of the 256 possible byte values. It is optimal in time but requires the storage of a 512-byte table. We begin by generating the table. The update of the CRC, for each byte of the message, is obtained by using the value read from the table;

– method without generating a table, where in the value of the table used in the previous method is calculated, on the fly. This method saves memory space for the table but is more costly in terms of resources and in time.

3.1.5. *Anti-collision protocol*

As most RFID systems operate in the master-slave mode, the RFID chip, as a slave, sends data only if it is requested by the reader who is the master. When multiple RFID tags are in the operating zone of a reader, the data received by the reader can interfere and become noise sources for the transmissions from their neighbours. The data received by the reader will then be wrong and we call it the data collision. An anti-collision process is a mechanism in which the reader is able to communicate with multiple RFID tags simultaneously present in its magnetic field. The anti-collision process ensures the integrity of the data transmission from multiple RFID tags to a single reader. The choice of various anti-collision protocols depends on:

– the intrinsic performance of the algorithms;

– the limitations of bandwidth;

– the implementation costs;

– the tolerance to noise;

– the integrity of signal;

– and the security.

For all these reasons, a vast majority of RFID technologies use the temporal distribution technique to implement the anti-collision process.

We distinguish deterministic and probabilistic algorithms. With the deterministic method we are able to determine the exact time for the collision management by counting the number of RFID tags present in the field, while with the probabilistic method we can estimate the probability to obtain the transponder IDs in a given time but without any guarantee.

3.1.5.1. *Deterministic anti-collision protocol*

The deterministic algorithms try to find out the unique identification number UID (*Unique Identification Number*) for each RFID tag, in an optimal time to select it if necessary [RAN 02].

One solution is to use a polling system which allows the reader to select a tag in its field from a list containing the tag identities. This requires a memory and a relatively small number of tags related to this reader. This system is very costly in terms of execution time even for a small number of tags present in the field. A much more common deterministic solution uses the Binary Tree Search Algorithm whose principle is to select a group of RFID tags, by identifying their UID, in an optimal and determined time. The choice of coding is important in the context of anti-collision process (Figure 3.5).

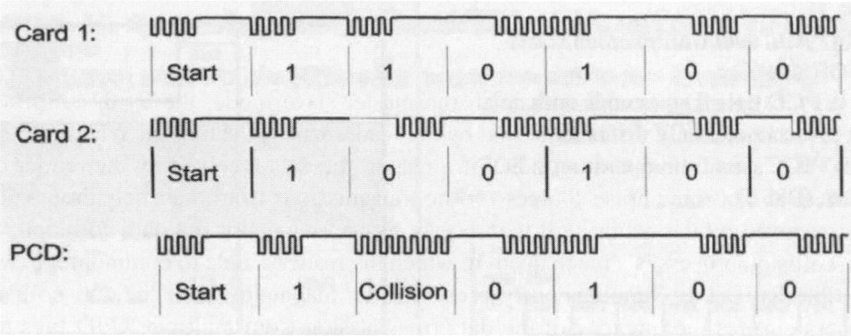

Figure 3.5. *Bit-oriented anti-collision*

In the first step, the reader performs a request to see whether the magnetic field has transponders capable of processing the collision management according to the mode proposed by it. At the end of this request, all transponders which can respond to this type of anti-collision process send, at a given time and in synchronism, a response to this request. If there is at least one transponder in the magnetic field, the next step is to see whether one or more transponders are simultaneously present in the field and individually separate them from each other by their unique identity code. If several transponders respond simultaneously and synchronously, because of their unique identifier, a collision occurs at least at the bit level, and then it is necessary to detect the collision. In this case, the Manchester coding plays its role as shown in Figure 3.1 where the detection of bit-oriented anti-collision presupposes that RFID tags respond synchronously at the request of the reader. At the time of the request, the reader sends a parameter to the tags where the parameter indicates the response delay between the end of a request and the beginning of the expected response and facilitates the synchronization process.

When a collision in a given bit is detected, the reader sends a request with the number of valid bits (the bits before the collision), which is immediately followed by a bit assigned to 1 (it can be 0, where this value is chosen by the designer of the reader).

Only those transponders which have the identifier part equal to the significant bits sent by the reader should be allowed to send the remaining bits of their identifiers. The anti-collision loop starts with several levels of cascading readings that depend on the size of the UID. The frames are then usually divided into two parts, one of which contains data derived from the reader to the RFID tag and the other has data from the RFID chip to the reader. The division between the two sides of exchanged frames depends on the position of the anti-collision loop. Moreover, the transmission direction between the reader and the RFID tag can be inverted when the desired number of bits has been sent. The anti-collision loop stops when the reader receives the frame from the RFID tag without any collision bit. The reader sends a selection command, usually with the UID of the RFID tag, followed by a CRC. The RFID tag, with the corresponding UID, confirms the command of the reader, usually with a SAK (SELECT-Acknowledge). According to its operating system or its state machine, it passes the tag from a reception/listening mode to an active mode to exchange and update data in EEPROM or RAM through a stable communication channel between the reader and the identified and selected RFID tag.

3.1.5.2. *Probabilistic anti-collision protocol*

For RFID systems in which the collision at the bit level is difficult to detect, we use the probabilistic protocol based on the principle of time slots by distributing randomly the transmissions of RFID tags in time. The reader sends a parameter to the RFID tags indicating the number of available slots (between 1 and N), where they can respond with a minimum of identification data. A collision is detected as a source of error by the CRC check in allowed time slots. The anti-collision process ends only when the communication from the RFID tag is controlled by the reader which allows, at an instant, only one tag to communicate without CRC error. The RFID tag/reader couple uses only the time slot which has been found to communicate during the session. Otherwise, the anti-collision process continues and the reader analyses the next time slot.

The response probability of RFID tag in each authorized slot is already defined. In addition, tags are authorized to dialog only in the framework of the anti-collision sequence initiated by the reader. Therefore, even if multiple RFID tags are in the field, the reader will find, with a probability, a time slot in which only one card has responded, and it will be selected to exchange data. However in the anti-collision process, other non selected RFID tags are put on hold (if the state machine allows them to do) in order to be able to dialog later with the reader [RAN 02].

The advantage of the probabilistic approach is that all RFID tags can be identified and selected using a single command, without having to contact each of them using their unique identification number.

The number N of slots is not fixed in advance. One can, in some cases, start the anti-collision loop with a relatively small number of slots and if many collisions occur, the number of slots will be increased in the next iteration.

3.2. Bibliography

[CAU 05] CAUCHETEUX D., BEIGNÉ E., RENAUDIN M., , CROCHON C., "Towards Asynchronous and High Data Rates Contactless Systems", *PRIME'05*, Lausanne, Switzerland, July 2005.

[RAN 02] RANKL W., EFFING W., *Smart Card Handbook, Third Edition*, Wiley, 2002.

PART TWO

RFID Applications

Chapter 4

Applications

4.1. Introduction

Nowadays, object identification and tracking are more and more advanced in their development. The globalization of the economy and the impact of new trade modes imply new challenges for logistics. A product should be tracked throughout its lifetime, i.e. from its production to destruction stage. In the first step, to ensure tracking, barcodes that enabled identification have been implemented. They have been considered as an essential tool for stock and flow management. However, during their usage step by step, limitations, mainly physical, have appeared:

– identification is required to pass by an optical reader (scanner);

– storage of some data.

As a second step, a new technology has been gradually developed. Qualified as an intelligent technology, it enables an automatic identification, in the same way as the barcodes. In addition to the fact that it allows a blind reading by radio waves, it can contain a large amount of information: this is RFID (*Radio Frequency IDentification*). These electronic tags, commonly indicated under the term RFID, will create an economic and cultural revolution comparable to the Internet in the 1990s. This technology will be used to recognize or identify, at a given distance, in the shortest time, an object, an animal or a person with a tag that is capable of exchanging data using radio waves. In our everyday life, we can mention, for example, the contactless smart card (NAVIGO, MONEO), the automatic tolling system (Liber-t), access controls to parking, etc. These inexpensive labels, which are usually produced in the billions, and can be remotely

Chapter written by François LECOCQ and Cyrille PÉPIN.

powered by electromagnetic fields, are gradually being integrated into objects used in everyday life. With the miniaturization and standardization (ISO in particular), the RFID technology has become well known thanks to innovative applications including distribution, transportation or industry. Indeed, it is an important technology to ensure the proper identification of any type of items. RFID can make data entry fast and automatic, using radio waves. Therefore it is increasingly used, especially in fields where other identification technologies, such as barcodes, face their own limitations. This new technology represents the beginning of the progress and their accession to the Internet of Things. However, it also faces criticisms about the privacy-related risks.

For easy reading, this chapter is divided into several parts:

– history: evolution from barcodes to RFID tags;

– RFID tags;

– normalization/standardization;

– advantages/disadvantages of RFID tags;

– description of RFID applications;

– application examples.

4.2. History: evolution from barcodes to RFID tags

4.2.1. *Description of barcodes*

Barcodes use different codification or symbolism protocols which depend on the constraints of utilization or standardization. They represent digital or alphanumeric data under the form of symbols constituted of bars and spaces with varying width (Figure 4.1). The bars can be replaced by small squares or dots, which correspond to two-dimensional codes. Barcodes were created in 1970 by George Laurer.

Figure 4.1. *Schematic representation of a barcode*

A barcode is a series of vertical lines of varying width, called bars and spaces. The set of bars and spaces is called "elements". There are various combinations of bars and spaces representing different characters. When reading a barcode, the light beam emitted by the scanner is absorbed by the dark bars without being reflected, while it is reflected by the clear spaces (Figure 4.2). Inside the scanner, a cellular photo detector receives the reflected light and converts it into an electrical signal. Thus, when an optical pen reads a barcode, the scanner creates a low electrical signal for the spaces (reflected light) and a strong electrical signal for the bars (nothing is reflected). The duration of the electrical signal determines if the elements are wide or narrow. The decoder of a barcode reader can convert this signal into characters represented by the barcode. The decoded data are then transmitted to the computer in a traditional format.

Figure 4.2. *Reading a barcode*

There are different types of barcodes, grouped into three main categories:
– one-dimensional (or linear) barcodes;
– stacked linear barcodes;
– two-dimensional barcodes.

In this part, different schemes and references were taken from [GOM].

4.2.2. *One-dimensional (or linear) barcodes*

In the category of one-dimensional barcodes, the main ones are:
– EAN (EAN-8, EAN-13, UPC) barcodes;
– codabar Monarch;
– code 11;
– code 39 and code 93;
– code 128;
– ITF (*Interleaved 2 of 5*) code.

4.2.2.1. *EAN (EAN-8, EAN-13, UPC) barcodes*

The EAN (*European Article Numbering*) number or code identifies objects in a unique manner. Under the form of barcodes, the number can be read by an optical scanner in an omnidirectional way.

The EAN code was developed from the US Code (*Universal Product Code*) for specific needs of the European trade. There are two variants, one of 8 digits (EAN-8) and the other of 13 digits (EAN-13), where the latter is the most common. They are represented under the form of sequences of black and white bars constituting a barcode (Figure 4.3).

Figure 4.3. *An example of EAN-8 code*

This code consists of a *BEGIN* zone (1 character), a data zone (8 (EAN-8) or 13 (EAN-13) characters) and an *END* zone (1 character) with the data area cut in the middle by a separator (1 character). A character in the data area has 2 bars and 2 spaces. Each bar or space has a width varying from 1 to 4 elements (constant measurement unit). Each character is composed of 7 white or black elements. There are two exceptions of this rule: the *BEGIN* and *END* zones which include a bar, a space and a bar (with the size of an element) are composed of 3 elements and the *SEPARATION* zone which includes a space, a bar, a space, a bar and a space (with the size of an element) is composed of 5 elements. The only difference between UPC and EAN-13 is that on UPC only 12 digits appear (instead of 13 digits on EAN). A UPC code is an EAN code in which the first two characters on the left are "0". In the UPC representation, only one of these zeros appears. The EAN code, like all barcodes, uses relatively specialized mathematical concepts. Their structure takes into account the physical constraints of the printing and reading conditions. Indeed, the recognition requires the ability to separate and measure the bar widths, at varying reading distances, on different types of sensors and in the absence of any clock or reference measurement. An EAN code is as follows:

– the prefix represents the code of the country that issued the participant number (2 or 3 digits) (30 to 37 for France);

– the participant number issued by the EAN organization in the concerned country (the following 4 or 5 digits);

– the item number of the object producer coded on 5 digits;

– the Check Digit calculated according to the 7 (EAN-8) or 12 (EAN-13) first digits which compose the code.

This type of barcode can be found on almost all the current products (food, clothing, stationery, household appliances, etc.). EAN-8 codes are reserved for small products (e.g. cigarette packets), while EAN-13 codes are used on all other products.

Figure 4.4. *An example of EAN-13 code*

4.2.2.2. *Codabar Monarch*

The Codabar Monarch is also called *NW-7* for 7 elements (bars and spaces) which are narrow or wide (Figure 4.5) and have one character.

Figure 4.5. *An example of Codabar Monarch*

The Codabar Monarch is not modular, where bars and spaces are irregular. The bar and space elements can have 18 different widths. It is auto-controlled, character by character:

– elements: 18 different widths (9 widths for bars and 9 widths for spaces);

– character: it consists of 7 elements (4 bars, 3 spaces). Among these 7 elements, 2 elements are wide and 5 elements are narrow;

– continuity: discontinuous, the spaces between characters are at least equivalent to the width of the most narrow element and at most, equivalent to the width of a character. They can vary even within a code;

– quiet zone: at least 100 mils (2.54 mm);

– set of characters: digits from 0 to 9 and 6 special characters ('-', '\$', ':', '/', '.', '+');

– *start/stop* characters: 4 characters, each of which can be represented by two letters ('a or t', 'b or n', 'c or *', 'd or e');

– Check Digit: this code is auto-controlled, it does not need the Check Digit.

Thanks to its relatively simple composition, Codabar Monarch is frequently used to encode serial numbers for applications in blood banks, home delivery services, etc.

4.2.2.3. *Code 11*

The name of the code 11 (or USD-8) is derived from the calculation method used to define its Check Digit.

The code 11 has a varying length (Figure 4.6). It allows us to codify 10 digits (0-9) and the dash character ("-"). Each character is composed of 5 elements (3 bars and 2 spaces). The characters are separated from each other by a narrow space.

Figure 4.6. *An example of code 11*

The code 11 is primarily used in the labelling of telecommunications equipment.

4.2.2.4. *Code 39 and code 93*

The code 39 has a varying length.

Figure 4.7. *An example of code 39*

Being alphanumeric, it can reference 26 capital letters of the Latin alphabet (A to Z), 10 digits (0 to 9) and 8 special characters ('-', '.', ' ', '*', '$', '/', '+', '%'). This code always begins and ends with the character "*" which is used as a trigger for the barcode reader. Each character is composed of 9 elements: 5 bars and 4 spaces. Each bar or space is "wide" or "narrow" and 3 of the 9 items are always "wide". It is also the origin of its name: code 39. The code 39 (Figure 4.7) is used in pharmacies, to reference drugs, in the automotive field, textile industry, etc. The code 93 (Figure 4.8) has been developed to improve the security and density of the code 39. This is an alphanumeric code of a varying length comprising a Check Digit on 2 characters "C" and "K". Like the code 39 from which it is derived, it can reference 26 capital letters of the Latin alphabet (A to Z), 10 digits (0 to 9) and 7 special characters ('-', '.', ' ', '$', '/', '+', '%'). The code 93 defines 5 additional special characters '!', '#', '&', '@', '*Start/Stop*'.

Figure 4.8. *An example of code 93*

A code 93 is always composed of the following elements: *Start/Stop*, data, *Check Digit* ("C"), *Check Digit* ("K"), *Start/Stop*. It is a rare code (used for postal services).

4.2.2.5. *Code 128*

The code 128 has a varying length and is a very dense code (Figure 4.9). It includes a *Start* zone (1 character), a data zone (12 characters) and a *Stop* zone (1 character). A character has 3 bars and 3 spaces. Each bar or space has a width varying from 1 to 4 modules (constant measurement unit). Each character is composed of 11 white or black modules, except the character *Stop*, which has 13. The code 128 is an alphanumeric code, which also implements the code for function keys (F1, F2, etc.), and 103 characters in the ASCII table.

Figure 4.9. *An example of code 128*

There are 3 sets of characters:

– A: upper-cases and Check Digits;

– B: upper-cases and lower-cases;

– C: digital (in pairs).

These 3 sets of characters can be mixed within the same code. It automatically includes a Check Digit. It is a common format in the industrial environment ("Sécurité Sociale" vignettes of French drugs, administrative forms, transport and logistic goods, etc.).

4.2.2.6. *ITF (or Interleaved 2 of 5) code*

The ITF (or *Interleaved 2 of 5*) code is of varying lengths and is only digital. The number of characters is *always* even, thus the code 123 is written as 0123. The term "interleaved" indicates the composition method of this code, i.e. data will be coded in pairs. The first data is encoded under the form of bars, while the second is in the form of spaces, thus bars and spaces are interleaved. There are 5 bars of which 2 are wide and 5 spaces of which 2 are large (Figure 4.10).

Figure 4.10. *An example of ITF (or Interleaved 2 of 5) code*

The code includes three variants (standard, interleaved and IATA). The IATA version is used for the labeling of baggages in airline.

4.2.3. *Stacked linear barcodes*

These barcodes are composed of several linear barcodes stacked on each other. They are read vertically by an automatic scanner reader (e.g. a camera), where the most common codes are:
– the PDF-417 code;
– the 16K code.

4.2.3.1. *PDF-417 code*

The PDF-417 code has a varying length (PDF stands for Portable Data File), which can include up to 1,850 alphanumeric characters or 2,710 numeric characters. This code allows us to print a large amount of information on a very small surface. The creation of a PDF-417 code occurs in 2 stages (Figure 4.11). The first step is to convert the data into codewords, which corresponds to a high-level coding. In the second step, these codewords are transcribed into a sequence of patterns of bars and spaces, which corresponds to a low-level coding. This code also has an error correction system that enables it to reconstruct the data which are wrongly printed, erased, fuzzy or torn. The PDF-417 code has the following main features:
– it contains more than 928 codewords for data;
– it is composed of 3 to 90 lines;
– a line contains 1 to 30 columns of data;
– there is a mechanism ('Macro PDF417') that allows distribution of data on several PDF-417 codes;
– there is an error correction level which varies from 0 to 8.

The large capacity of the PDF-417 code is used when detailed information must always be attached to the identified object, such as the case in the transportation of dangerous materials for example.

Figure 4.11. *An example of PDF-417 code*

4.2.3.2. *Code 16K*

The code 16K was developed in 1989 by Ted Williams, who also developed the code 128. The code 16K structure is based on the code 128.

The main features of the code 16K are:

– it has a varying length;

– it allows the codification of the first 128 ASCII characters;

– its maximum density is 32 alphanumeric characters or 65 numeric characters per cm^2;

– it has 2 to 16 lines of 5 ASCII characters.

The 16K code (Figure 4.12) is used in many fields such as defense, health, electronics and chemical industry, etc.

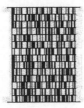

Figure 4.12. *An example of code 16K*

4.2.4. *Two-dimensional barcodes*

These barcodes are the most complex and elaborate. They are horizontally encoded *and* read. Therefore they can encode more data on the same surface. The most common two-dimensional barcodes are:

– code One;

– Aztec code;

– DataMatrix code;

– MaxiCode;

– QR code.

4.2.4.1. *The code One*

The code One was invented in 1992 by Ted Williams (who is also the inventor of the code 128, code 16K). Its features are:

– its varying length;

– its capacity is 2,218 alphanumeric characters or 3,550 numeric characters;

– its capacity to print lot of information on a very small surface.

This code (Figure 4.13) is rarely used. It can be found mainly in electronic and chemical industries.

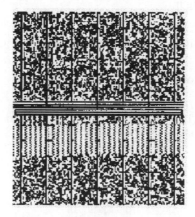

Figure 4.13. *An example of code One*

4.2.4.2. *Aztec code*

The Aztec code was invented in 1995 by Andy Longacre. Its features are:

– its varying length;

– its capacity to encode up to 3,750 ASCII characters.

A symbol is built around a square target located in the center. The data is codified in a series of layers surrounding the target. Each additional layer surrounds the previous layer, which implies that the symbol remains square during the progressive data addition.

Figure 4.14. *An example of code Aztec*

The Aztec code (Figure 4.14) can have a multitude of sizes for coding small or large amounts of data. This code can be read regardless of its orientation and contains an error correction mechanism (these error correction codes are used in DVDs, mobile phones, satellites, ADSL modems, etc.). The smallest element of an Aztec code is called a module which is a square point. Aztec code size depends on the size of the module, the amount of data and the error correction level chosen by the user. The smallest Aztec Code has 15 side modules and can hold up to 14 characters with a level of error correction of 40%. The biggest symbol contains 151 side modules and holds 3750 characters with a level of error correction of 25%. The Aztec Code is primarily used for transportation. For example, Deutsche Bahn, the German national railway company, uses this type of code which is sent to a telephone as a ticket. During a control operation, the controller will scan the code on the phone screen.

4.2.4.3. *DataMatrix code*

The DataMatrix code is presented in the form of a matrix of juxtaposed dots or squares. Its main features are:

– its varying length;

– its capacity of 2,335 alphanumeric characters or 3,116 numeric characters;

– its capacity to print lot of information on a very small surface;

– its system for reading errors correction.

The DataMatrix code (Figure 4.15) can contain different security levels enabling it to be read even when it is partially degraded or obscured. The higher the security level, the larger the symbol size. There are several variants of the DataMatrix code accepted

Figure 4.15. *An example of DataMatrix code*

by the standard: from ECC000 (ECC pour *Error Check Correction*) which provides no security if the symbol is degraded, like one-dimensional barcodes (e.g. EAN 13), to ECC200 which provides the maximum security level (a symbol reading is still possible even if it is obscured up to about 20%, i.e. 80% intact). The DataMatrix code is used in electronics industry for marking printed circuit boards and integrated circuits.

4.2.4.4. *MaxiCode*

The MaxiCode, which is a two-dimensional barcode, was invented by the company UPS for its packages in 1992.

Figure 4.16. *An example of MaxiCode*

The MaxiCode (Figure 4.16) allows the representation of 100 alphanumeric characters on a surface of 6.45 cm^2. Each MaxiCode consists of a central element, and is similar to a target, surrounded by a square matrix of 33 rows of 29 or 30 (alternatively) hexagonal elements. The MaxiCode is mainly used for transportation (especially by UPS).

4.2.4.5. *QR code*

The QR code is a two-dimensional code that was developed by a Japanese automotive supplier in 1994. QR stands for Quick Response.

Figure 4.17. *An example of QR code*

The QR code (Figure 4.17) is of varying length and can include up to 4,296 alphanumeric characters or 7,089 digits. This code stores a lot of information while maintaining a small size and is easy to scan. This code is mainly used in Japan, especially for mobile telephones.

4.3. RFID tags

The barcode is envisaged technology in the field of identification. However, as we have seen, this technology is particularly limited to data storage applications, thus resulting in the emergence of a new technology: RFID tags. The radio-identification, usually known by the acronym RFID, is a technology used to store and retrieve remote data using markers called "radio-tags" ("RFID tag" or "RFID transponder"). RFID is presented in the form of small objects, such as self-adhesive labels which can be pasted or embedded in objects or products and even implanted in live organisms (animals, human body).

RFIDs have an antenna associated with an electronic chip that enables them to receive and respond to requests issued by a radio transceiver. This recognition technology can be used to identify:

– objects as with the barcode technology, in this case, they are called electronic tags;

– people, when a RFID is integrated in a passport, a travel card, a debit card. In this case, they correspond to contactless cards.

4.3.1. *Characteristics of RFID tags*

RFIDs are commonly called tags but they are also found under the name smart tag or transponder (equipment for receiving a radio signal and returning immediately another radio signal containing a relevant information). In a conceptual side, RFID tags and barcodes are quite similar, both are designed to provide a rapid and reliable identification of elements and possibilities of filiation. The main difference between these two technologies is that the barcode is read with an optical laser, while an RFID tag is scanned by a reader that retrieves radio frequency signals emitted by the tag. The main characteristics of a RFID are:

– a large storage capacity (which defines the number of characters that can be encoded): 1 to several kilobytes (KB);

– information can be read by a scanner at a distance varying from few centimeters to about 200 meters;

– an RFID does not require any contact or particular vision field to operate;

– an RFID can operate in multiple environments (water, darkness);

– depending on the type of RFID, it is possible to read but not necessarily to write on the RFID, reading/writing operations are performed in contactless mode. On some RFIDs, it is also possible to re-register information, then recycle the tag;

– it is easy to use and is quite suitable for automatic processing.

4.3.2. *Operating principle*

The RFID technology uses radio frequencies between 50 kHz and 2.5 GHz. A RFID system (Figure 4.18) is composed of the following elements:

– an integrated circuit (a microchip), which contains the data of the element to identify;

– an antenna used to transmit signals between the reader and the RFID tag;

– a reader, which receives signals from the RFID and is responsible for processing.

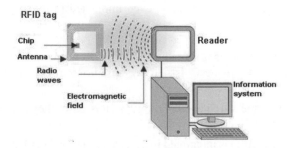

Figure 4.18. *RFID system*

Today, the size of the tag can be reduced to the size of a point (≈ 1 mm^2). It also presents problems of packaging (sawing is performed using a laser rather than a mechanical method). The antenna (Figure 4.19), often composed of copper, is embedded in the tag using ultrasound (vibration systems). The antenna will enable the RF (radio frequency) transmission or reception. In some systems, a single antenna performs two operations while in other systems, an antenna sends signals and the other one receives signals. The antenna is a fundamental element in a radio-electric system and its characteristics related to performance, gain, radiation diagram influence the quality and the scope of the system directly. An important point to consider is the interaction between the reader and RFID.

Figure 4.19. *Antenna – RFID*

There are then two couplings:

– inductive or magnetic coupling;

– radiative or electromagnetic coupling;

4.3.2.1. *Inductive or magnetic or near field coupling*

One of the basic principles of antennas is the radiation emission (Figure 4.20).

– B: magnetic field;

– μ: magnetic permeability;

– H: magnetic excitation.

When a tag enters into the reader field, it can access data stored in the RFID. The data reading is done by analyzing the perturbations induced in the emitted field. There is a limitation in term of scope and the magnetic field decreases with the use of low frequencies. The frequency of 13.56 MHz (high frequency or HF) corresponds to a wavelength of 22 meters. The UHF (Ultra High Frequency) and SHF (Supra High Frequency), which correspond to wavelengths ranging from meter to centimeter, cannot be used in the near field. These frequencies will enable the radiative coupling. RFID systems in magnetic coupling generally use passive tags. The element for communication between the tag and the reader is a set of several metal coils that produce the energy needed to power the embedded electronic of the tag. It exploits the induction phenomena created by the magnetic field emitted by the reader. This technique will introduce a number of constraints: the limited communication distance

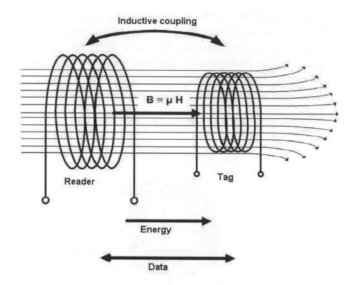

Figure 4.20. *Inductive or magnetic coupling*

due to the emission of the magnetic field near the reader (less than one meter) and the generation of perturbations to other systems near the reader. The operating mode in magnetic coupling concerns the systems operating at low frequency (from 120 kHz to 135 kHz) and at high frequency (13.6 MHz). Magnetic coupling systems operate up to 29 MHz. Historically, the first magnetic systems operated at low frequencies, followed by the emergence of the first high-frequency applications. This type of tag (antenna and circuit/memory) is usually engraved on a flexible substrate of a size less than 10 cm, and is integrable on devices such as smart cards.

4.3.2.2. *Radiative or electromagnetic or far field coupling*

In far field, at a distance approximately greater than the wavelength from the source, the beam diverges to form a spherical wave which is locally plane. The RFID behaves like a real radio transceiver and generally requires active solutions:

– x, y, z: coordinates along the 3 axes;

– H: magnetic field;

– E: electric field;

– d: distance.

The electromagnetic field decreases with $\frac{1}{d}$, inverse of the distance to the source and received energy decreases with $\frac{1}{d^2}$. This interaction mode allows communication over larger distances, which is more than 10 meters, and data transmission at higher rates. Antennas, which depend on the wavelength, are smaller. Radiative systems

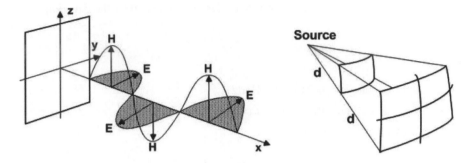

Figure 4.21. *Radiative or electromagnetic coupling*

(Figure 4.21) are more complex than inductive systems. The wave propagation is more difficult to predict, sometimes random, and interference phenomena are more difficult to handle. However, operating at 900 MHz, it is possible to significantly reduce the antenna size which contributes to the miniaturization of equipments. Contrary to magnetic coupling modules, the systems in electric coupling are not constrained by the emission that is localized around the reader of field lines. Using propagation properties of the electromagnetic field radiated by an antenna, it is possible to transport energy and data from one reader to a transponder and vice versa at more than 10 meters. The dimensions of antennas capable of producing such electric fields are of the order of half wavelength (for a frequency of 100 MHz, the antenna size is about 1.50 m). In practice, it is necessary to carefully select the transmission frequency of the antenna (UHF should be preferentially used), so that the restriction of electric coupling in RFID is as low as possible. In the case of passive transponders, the supply energy is created by exploiting the Hertz dipole phenomenon produced by the electric field emitted by the reader. The energy density of radiated signal decreases with the inverse square of the distance between the reader and the tag. Thus, passive systems can be used only at distances up to 10 meters for frequencies around 500 MHz. This distance decreases sharply when the frequency increases (less than one meter at 2.5 GHz). Beyond these frequencies, transponders require power and become active. A major advantage of tags in electromagnetic coupling is certainly its low production cost due to development techniques used by microelectronic designers.

The temporal or momentary frequency f is the inverse of the period T which is the time unit:

$$f = \frac{1}{T}$$

If the selected time unit is second (symbol s), the frequency is then measured in hertz (symbol Hz), in the international system of units. The higher the value in hertz, the shorter the duration in seconds.

The frequency of a propagated wave can also be calculated by the equation:

$$f = \frac{c}{\lambda}$$

with:

– f: the frequency of the wave (in Hertz);

– c: the speed of the wave (in meter per second);

– λ: the wavelength (in meter).

A radio signal emits in a range of some hundred meters, depending on the power of the installation, and especially its frequency. A radio wave is classified according to its frequency in Hertz or cycles per second. All of these frequencies form the radio frequency spectrum:

– Low Frequency (LF): frequency from 30 to 300 kHz with a wavelength of 1 to 10 km and a reading range of 10 to 50 cm;

– Medium Frequency (MF): frequency from 300 kHz to 3 MHz with a wavelength of 100 m to 1 km and a reading range of 50 to 80 cm;

– High Frequency (HF): frequency from 3 to 30 MHz with a wavelength of 10 to 100 m and a reading distance of 5 cm to 3 m;

– Ultra High Frequency (UHF): frequency from 300 MHz to 3 GHz with a wavelength of 10 cm to 1 m and a reading distance of 1 to 5 m;

– Super High Frequency (SHF): frequency from 3 GHz to 30 GHz with a wavelength of 1 to 10 cm and a reading range higher than 10 m (it can be reduced by the presence of metal).

For low, medium and high frequencies, the coupling between the tag and the reader is the inductive coupling (or near field). For ultra and super high frequencies, the radiative coupling (or far field) is applied.

Waveband	Frequency	Wavelength
Low Frequency (LF)	30 to 300 kHz	1 to 10 km
Medium Frequency (MF)	300 kHz to 3 MHz	100 m to 1 km
High Frequency (HF)	3 to 30 MHz	10 to 100 m
Ultra High Frequency (UHF)	300 MHz to 3 GHz	10 cm to 1 m
Super High Frequency (SHF)	3 to 30 GHz	1 to 10 cm

Table 4.1. *Frequencies/wavelengths*

Like all electromagnetic waves, radio waves propagate through empty space at the speed of light and with an attenuation proportional to the square of the travelled distance. In the atmosphere, they undergo further attenuations related to precipitation. They are attenuated or deviated by obstacles, depending on their wavelength, the nature of the material, its shape and size. For simplicity, a conductive material will have a reflection effect, while a dielectric material will produce a deviation. In conclusion, the effect is related to the ratio between the size of the object and the wavelength.

4.4. Normalization/standardization

Real success of the RFID technology lies in overcoming a major challenge posed by standardization. According to its current expansion, it will invade all fields like computers did. The current trend of companies is to operate in connected collaboration systems. In order to enable the identification of a product or a range outside the company along a supply chain with multiple entities, it is necessary to deal with standardization. There are many advantages to set established standards for a precise frequency range. In 1999, with the collaboration of several companies, the Auto-ID Center was established at the initiative of *MIT* (Massachusetts Institute of Technology). The mission of the center was to solve, among others, standardization constraints. This initiative was subsequently taken over by EPCglobal, an initiative supported by *Electronic Article Number* (EAN) and UCC to finalize the standardization of the RFID technology.

Waveband	Utilization	Notes	Protocols
Low Frequency LF	Animal identification	No influence in the presence water or metal Transfer rate: < 10 kb/s Reading distance: 10 to 50 cm	ISO/IEC 18000-2 ISO 10536
Medium Frequency MF	Contactless payment	No influence in the presence water or metal Transfer rate: < 50 kb/s Reading distance: 50 to 80 cm	ISO 14443 (type A & B & C) ISO 10373-6
High Frequency HF	Access control	Reading perturbation in the presence of water or metal Transfer rate: < 100 kb/s Reading distance: 5 cm to 3 m	ISO/IEC 18000-3 ISO 15693 ISO 10373-7 EPC HF
Ultra High Frequency UHF	Range counting	Reading perturbation in the presence of water or metal Transfer rate: < 200 kb/s Reading distance: 1 to 5 m	ISO/IEC 18000-6 ISO/IEC 18000-7 EPC Class 0/0+ EPC Class 1 Gen1 EPC Class 1 Gen2
Super High Frequency SHF	Vehicle identification	Reading perturbation in the presence of water or metal Transfer rate: < 200 kb/s Reading distance: ≈ 10 m	ISO/IEC 18000-4

Table 4.2. *RFID application fields based on waveband*

EPCglobal has developed a software that plays the role of a nervous system of technology networks. This system promotes the use of the ONS (*Object Naming Service*) for the identification of an EPC via a local network or Internet to look for the associated product. The ONS directs the software to a database of companies that stores information about products. All information across all networks will be stored and exchanged within information systems in a new language (*Physical Markup Language*) based on XML (*Extensible Markup Language*).

The communication by radio frequencies between the tag and the reader is defined by the protocol standardized by ISO. In reality, the ISO 18000 standards, presented as *the* solution to the interoperability problem, are not sufficient to achieve this goal. Two conditions must be fulfilled: on the one hand a common communication protocol between reader and tag should be used, since it is the role of ISO 18000 standards, and on the other hand, the structure of data contained in tags should be organized in a unique way. The ISO 18000 standards are part of a group of standards which, taken together, allow interoperability.

The communication protocol is the language used by readers and tags to communicate. Like all languages, there is a vocabulary and syntax covering the commands and data contained in tags.

4.4.1. *ISO standards for RFID*

The first standards are the ISO 18000 protocol series:

– ISO 18000-1: Radio frequency identification for item management: Reference architecture and definition of parameters to be standardized;

– ISO 18000-2: Radio frequency identification for item management: Parameters for air interface communications below 135 kHz;

– ISO 18000-3: Radio frequency identification for item management: Parameters for air interface communications at 13.56 MHz;

– ISO 18000-4: Radio frequency identification for item management: Parameters for air interface communications at 2.45 GHz;

– ISO 18000-5: Radio frequency identification for item management: Parameters for air interface communications at 5.8 GHz (abandoned);

– ISO 18000-6: Radio frequency identification for item management: Parameters for air interface communications at 860 MHz to 960 MHz;

– ISO 18000-7: Radio frequency identification for item management: Parameters for active air interface communications at 433 MHz.

Notes related to the ISO 18000 standards:

Note on ISO 18000-3: 2 modes are used: the mode 1 derived from ISO 15693 for contactless cards and the mode 2 derived from the technology developed by Magellan

(Australia). Its characteristic is to allow a much faster (up to 40 times) data exchange speed. Note that these two modes are not interoperable.

Note related to ISO 18000-6: 3 types are used:

– type A uses the Pulse Interval Encoding (PIE) system with slotted ALOHA collision arbitration protocol;

– type B uses the Manchester Encoding system with Binary Tree collision arbitration protocol;

– type C is based on the proposition of EPCglobal Class1 Gen2.

These three types (A, B and C) are not interoperable.

In parallel with the definition of the ISO 18000 standards in different frequencies, the corresponding test methods were implemented (these methods are attached to compliance tests):

– ISO 18047-2: Radio frequency identification device conformance test methods: Test methods for air interface communications below 135 kHz;

– ISO 18047-3: Radio frequency identification device conformance test methods: Test methods for air interface communications at 13.56 MHz;

– ISO 18047-4: Radio frequency identification device conformance test methods: Test methods for air interface communications at 2.45 GHz;

– ISO 18047-5: Radio frequency identification device conformance test methods: Test methods for air interface communications at 5.8 GHz;

– ISO 18047-6: Radio frequency identification device conformance test methods: Test methods for air interface communications at 860 MHz to 960 MHz;

– ISO 18047-7: Radio frequency identification device conformance test methods: Test methods for active air interface communications at 433 MHz.

The main ISO standards for contactless RFID are the ISO 14443 standards (*contactless* mode at 10 cm) with power consumption up to about 10 mW, operating distances below 10 cm, a data rate of several hundred kilobits per second and fields around 5 A/m. The ISO 14443 series includes:

– ISO 14443-1: Contactless integrated circuit cards: Physical characteristics;

– ISO 14443-2: Contactless integrated circuit cards: Radio frequency power and signal interface;

– ISO 14443-3: Contactless integrated circuit cards: Initialization and anti-collision;

– ISO 14443-4: Contactless integrated circuit cards: Transmission protocol.

Regarding the characteristics of close coupled cards (*close* mode at 10 mm), they are defined in the ISO 10536 standards as follows:

– ISO 10536-1: Contactless integrated circuit(s) cards: Close-coupled cards: Physical characteristics;

– ISO 10536-2: Contactless integrated circuit(s) cards: Close-coupled cards: Dimensions and location of coupling areas;

– ISO 10536-3: Contactless integrated circuit(s) cards: Close-coupled cards: Electronic signals and reset procedures.

There are also standards for vicinity mode (1 m), namely:

– ISO 15693-1: Contactless integrated circuit(s) cards: Vicinity cards: Physical characteristics;

– ISO 15693-2: Contactless integrated circuit(s) cards: Vicinity cards: Air interface and initialization;

– ISO 15693-3: Contactless integrated circuit(s) cards: Vicinity cards: Anti-collision and transmission protocol.

Like ISO 18000 standards, test methods have been developed for proximity fields and vicinity fields:

– ISO 10373-6: Test methods: Proximity cards;

– ISO 10373-7: Test methods: Vicinity cards.

Standards have also been specified to measure performances:

– ISO 18046-1: Automatic Identification and Data Capture Techniques: RFID Performance: Test Methods for RFID systems;

– ISO 18046-2: Automatic Identification and Data Capture Techniques: RFID Performance: Test Methods for RFID interrogators;

– ISO 18046-3: Automatic Identification and Data Capture Techniques: RFID Performance: Test Methods for RFID tags.

Standards related to radio frequency identification are:

– ISO 15961-1: Radio frequency identification (RFID) for item management: Data protocol: Application interface;

– ISO 15961-2: Radio frequency identification (RFID) for item management: Data protocol: Registration of RFID data constructs;

– ISO 15961-3: Radio frequency identification (RFID) for item management: Data protocol: RFID data constructs;

– ISO 15961-4: Automatic Identification and Data Capture Techniques: Radio Frequency Identification (RFID) for Item Management: Application interface commands for battery assist and sensor functionality;

– ISO 15962: Radio frequency identification (RFID) for item management: Data protocol: Data encoding rules and logical memory functions;

– ISO 15963: Radio frequency identification for item management: Unique identification for RF tags;

– ISO 19762-3: Automatic identification and data capture (AIDC) techniques Harmonized vocabulary: Radio frequency identification (RFID);

– ISO 24753: Automatic Identification and Data Capture Techniques: RFID for item Management: Application Protocol: Encoding and processing rules for sensors and batteries.

Using these standards allows "RFID solution integrators" to find the systems adapted to the needs of their clients based on the verified performances. They can constitute a reference even if the tests are not performed on site. It will also allow users to make a choice among several options. Indeed, the interoperability allows us to suggest that any reader compliant with ISO 18000 can read any RFID tag which is also compliant with the same standard. But the interoperability does not mean that all systems on the market have the same performance where all things are equal. In all cases, the capture of information will certainly be guaranteed, which is not guaranteed in cases related to the distance and reading speed, or even reading rates in a given electromagnetic environment.

4.4.2. *ISO standards for middleware*

For middleware, there are also standards for managing the exchange:

– ISO 24791-1: Automatic Identification and Data Capture Techniques: RFID for item Management: System Software Infrastructure - Architecture;

– ISO 24791-2: Automatic Identification and Data Capture Techniques: RFID for item Management: System Software Infrastructure - Data Management;

– ISO 24791-3: Automatic Identification and Data Capture Techniques: RFID for item Management: System Software Infrastructure - Device Management;

– ISO 24791-4: Automatic Identification and Data Capture Techniques: RFID for item Management: System Software Infrastructure - Application Interface;

– ISO 24791-5: Automatic Identification and Data Capture Techniques: RFID for item Management: System Software Infrastructure - Device Interface;

– ISO 24791-6: Automatic Identification and Data Capture Techniques: RFID for item Management: System Software Infrastructure - Security.

4.4.3. *User guidance*

Related to these new technologies, standards or, more generally, user guidances have been developed for the RFID management. The main guidelines are:

– ISO 24729-1: RFID - Enabled Labels;

– ISO 24729-2: Recycling and RFID Tags;

– ISO 24729-3: Implementation and operation of UHF RFID Interrogator system in Logistics Applications.

All references and additional explanations on ISO standards are found on the site [ISO].

4.4.4. *Protocols*

It is certainly obvious that the RFID technology replaces barcodes, and the protocols used are continuously improving. In general, the implementation of the RFID technology uses the EPC Generation 2.0 protocol as standard. Based on their features, the center of automatic identifications distributes tags into six classes from Class 0 to Class 5. The tags used for identification are generally less expensive and are in the Class 0 and Class 1 types. Knowing that identification is the most used function, identification tags are the most popular and, more importantly, their use is considered to have changed the RFID technology.

4.4.5. *EPCglobal standards*

The amount of existing solutions made the universal traceability difficult to perform. The EPCglobal consortium [EPC] is an organization working on an international standard proposition to standardize RFID techniques. The goal is to have a homogeneous distribution system of identifiers in order to have an EPC (*Electronic Product Code*) code for each item in the logistic chain of each company in the world. Like the barcode, more than 30 years ago, EAN.UCC and companies have defined a standardized solution for the object identification via RFID. Today, the first generation of the EPC standard is available. This result is due to research efforts, funded by hundreds of companies from all countries. Gathered within the AutoID Center, the research involves 5 of world's leading universities, Massachusetts Institute of Technology (MIT), University of Cambridge in Great Britain, University of St. Gallen in Switzerland, University of Adelaide in Australia and University of Keio in Japan. Together, they defined components and built the base of the future "Internet of Things". EPCglobal defines an end-to-end tag management model, relying on the Internet and the ONS (*Object Naming Service*) specification, inspired from the (*Domain Name Service*) model. Six classes of tags are proposed:

– class 0: read only (passive);

– class 1: write once, read multiple;

– class 2: rewritable (passive);

– class 3: rewritable (semi-passive);

– class 4: rewritable (active);

– class 5: active readers (class 4 with possibilities similar to a reader to communicate with other tags).

Therefore the EPC system (Figure 4.22) tends to become a global, modular and interoperable architecture, which can articulate key features of future information exchanges: unitary monitoring of objects through the EPC, remote information capture with RFID, storage and access to information through open standards of the Internet.

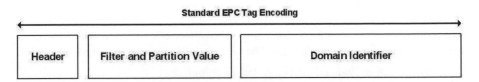

Figure 4.22. *Standard representation of an EPC tag*

The EPC provides a serial number based on the current product code for a unique identification of products. Tags offer, at reasonable price, in a satisfying cost, a sufficient storage capacity to store new information. When they are coupled in an antenna and read using radio frequency technology, distance reading becomes a reality. Finally, with the help of the Internet technology, information systems of EPC can store, communicate and allow a shared and controlled access to information for all participants in the chain (Figure 4.23).

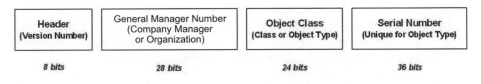

Figure 4.23. *EPC Gen 1 representation*

Initially, the General Identifier EPC of 96 bits (GID 96) has been established. It is commonly used.

4.4.6. *Communication layer*

The dialog between tag and reader has constraints related to the RFID technology. The goal of the communication layer is to define the rules and ensure its communication is effective. It manages:

– recognition and identification of one or more tags in the field of the reader;

– anti-collision algorithm to interact with several tags in the same field;

– presentation of data in the tag;

– memory size of the tag;

– reading all or a part of the tag content;

– writing on the tag to change the content or add information;

– security of an exchange;

– end of a dialog.

These different elements must be standardized to allow any reader to interact with any tag, regardless of their manufacturer. The security of an exchange includes:

– integrity of exchanged data (e.g. data encryption);

– authorization to read or write;

– protection of data;

– protection during data transfer (e.g. data bus is encrypted).

4.4.7. *Different types of tags*

A tag is composed of the following elements:

– an antenna;

– a silicon chip;

– a substrate and/or an encapsulation.

A tag is very discrete by its thinness, its small size (few millimeters), and its mass which can be considered negligible. Since its cost is now minimal, we can consider it is as disposable, although its re-use is more environmentally friendly.

The tags are divided into four groups:

– passive tags;

– semi-passive tags;

– semi-active tags;

– active tags.

"Semi-passive", "active" and "semi-active" tags require a battery (they are also called BAP, *Battery-Assisted Passive tags*).

4.4.7.1. *Passive tags*

Passive tags require no power source other than the one provided by the reader. Earlier, the reading operation of passive tags was limited to a distance of about 10 meters, but now, thanks to technologies used in communications systems, this distance can extend up to 200 meters. The reader sends energy to the tag, along with an interrogation signal that the tag responds to. One of the possible simplest answers is the return of a digital ID, such as the one of the EPC-96 standard using 96 bits (Figure 4.23). Following which a table or a database can be consulted to ensure an access control, a statement or a data tracking, and any expected statistic. Passive tags operate in read-only mode (like linear barcodes). In this case, the antenna captures some frequencies which provide enough energy to issue its unique identification code. The passive tags are programmed with data which is not modifiable, with a capacity of 32 to 128 bits. They are supplied blank for users. In most cases, the provider gives them with an identification. During its installation on an object, the user will provide the data which will be useful for him. During the later life of the tag, this information can be read but cannot be modified or completed. Further sophisticated devices have sensors to enable them to identify the physical variations such as temperature (e.g. for frozen products). Some tests were made with magnetic ink that acts as an antenna. Passive tags are inexpensive and have an almost unlimited lifetime. They constitute the bulk of the market, and further more in the future for logistic applications in open circuit, where the tag is lost during the sale of a product. Indeed, these tags are deactivated onca a product has been purchased. Currently, the most common passive tags are EPCs (*Electronic Product Code*) with the following features:

– operating at 13.56 MHz frequency;

– operating in read-only mode;

– coding on 96 bits.

These EPCs represent a new product that can be used to detect, track and control a variety of items using the RFID technology. The EPC structure can distinguish single items of the same type. For example, two DVDs (corresponding to the same video) have the same standard universal product code (UPC: this code is strictly reserved for barcodes). This allows computer systems to determine the manufacturer of the DVD, the movie title and facilitates trade or sale. EPC extends the UPC code and ensures that two videos of the same type are distract from another one anther: each item is individually identified.

4.4.7.2. *Semi-passive tags*

Semi-passive tags require a battery (rechargeable or not) to perform calculations or use internal sensors. It is the reader that provides the necessary energy for communications with these tags.

4.4.7.3. *Semi-active tags*

Semi-active tags do not use their battery to transmit signals. They act as passive tags at the communication level. However the battery allows them, for example, to store data during transport. These tags are used for the product shipments under a controlled temperature and to record the temperature of goods at regular intervals. These tags stay in the idle mode until they receive an activation message from the reader.

4.4.7.4. *Active tags*

Active tags are composed of:

– a battery (rechargeable or not) allowing them to emit a signal;

– a processor;

– a memory.

Contrary to passive tags, active tags can be read at long distances. However, an active information transmission signalizes the presence of tag(s) to all and thus raises problems about the security of goods. Active tags are powered by an internal extra flat battery and allow data reading and writing, with a memory of up to 10 kilobits.

4.5. Advantages/disadvantages of RFID tags

4.5.1. *Advantages*

Even now, barcodes remain as the principal automatic identification technology used in supply chains. Therefore, barcode is considered as a reference to analyse the advantages and disadvantages of RFID.

The advantages of radio frequency tags compared to barcodes are:

– possibility to update information contained in an RFID;

– greater capacity for storing information;

– higher data recording speed;

– increased security for data access;

– greater flexibility in tag positioning;

– higher lifetime especially in terms of scalability;

– better protection against environmental conditions.

4.5.1.1. *Update*

Unlike barcodes for which data are frozen when they are printed or marked, data stored in an RFID tag can be modified (the amount of information stored on the tag can be increased or decreased) by authorized persons. Therefore, they must be accessible in reading and writing operations.

4.5.1.2. *Storage*

Gradually during the technology development, barcodes have appeared for storing large amounts of data especially with two-dimensional barcodes. However, their use in the industrial and logistic world remains problematic because they require special conditions for printing and/or reading. Most widely used barcodes are limited to data contents of less than 50 characters, and in extreme cases require a large label (about an A4 or A5 sheet).

In an RFID tag, the storage of 1,000 characters can be done easily on an area less than 1 mm^2. It can even reach 10,000 characters. For example, in a logistic tag affixed to a lot, different elements and their respective quantities can be recorded and read. The deployment of this technology allows the suppression of physical inventories (exact tracking and localization of products). It can also record all transactions performed on a finished product.

It has been observed that there is a reduction of input errors or transmissions, and a reduction of reaction time when new information is acquired (e.g. trigger a restocking when the number of products on shelf falls below a given threshold).

4.5.1.3. *Recording speed*

Barcodes in a logistic context often require hard copy printing. Tag handling and placing are still manual or mechanical operations. RF tags can be included in the handling support or the initial packaging stage. Data related to objects is rapidly written at the constitution of the logistic or transport unit, without additional manipulation.

4.5.1.4. *Access security*

Like any digital medium, an RFID tag can be protected by password for reading or writing. Data can be encrypted. In the same tag, a part of information can be freely available and the others protected. These options allow the tag to be a good option against theft and/or counterfeiting. It is also possible to simultaneously differentiate several objects with an anti-collision system, which can handle a larger number of objects in the shortest possibe time.

4.5.1.5. *Positioning a tag*

To enable the automation of barcode reading, the standardization bodies have defined positioning rules of logistic units. With an RFID tag, it is possible to abstract the constraints related to the optical reading: as it does not need to be seen, the tag can be found under the packaging and even inside the product. It needs only to enter into the field of the reader in order that its presence is detected. Data reading/writing are performed "on the fly", which leads to a greater positioning and flexibility of the tag as its detection is automatically done. Most of the time, the tag has a small size, about the size of a stamp.

4.5.1.6. *Life*

In applications where the same object can be used several times, such as the identification of handling supports or the consignment of container, these tags can have a life span higher than 10 years. They are supplied blank and can be modified several times, or deleted and read. The number of repetitions of these operations can be more than 500,000 or a million times, which gives them a high reliability and very interesting value for money.

4.5.1.7. *Environmental conditions*

RFID tags do not need to be positioned outside the object to be identified. Therefore, they may be better protected from external aggressions related to storage, handling or transport. In addition, their operating principle does not make them sensitive to various tasks that may affect the use of barcodes. These tags can be more resistant to external aggressions as they can be covered with a packaging, under the condition that the latter is not an obstacle of reading/writing operations of the tag.

4.5.2. *Disadvantages*

There is no doubt that this technology has many advantages as described in the preceding section. However, RF tags also have constraints, the main ones are:
– particularly fluctuating cost;
– disturbance due to the physical environment;
– perturbations induced by tags among them;
– sensitivity to interference of magnetic and electric waves;
– unprintable;
– regulation constraints related to the impact on health.

4.5.2.1. *Cost*

The price of radio frequency tags will depend on many factors:
– size of memory;
– type of packaging, assembly;
– technology;
– features (especially security);
– volume.

The price is clearly higher than barcode's one for consumer units. Indeed, an EAN/UPC barcode included in the packaging of a product costs nothing, a self-adhesive tag of the same type is less than 3 cents if it is passive but up to 25 euros if it is active. Today, using radio frequency tags instead of barcodes on consumer products is not economically feasible. However it becomes necessary to counter against theft or counterfeiting on high value added products or to follow products in the context of after-sale service, such as home appliances or hi-fis. However, beyond the unitary packaging, the cost of an RFID tag can be marginal compared to the value of contained products. Therefore in the field of consumer products, the initial applications of these tags can be created on boxes, pallets and transport units. Moreover, if the comparison is done at the level of identification and tracking system, we must take into account the costs of readers, favorable to RFID, and time saving due to the non obligation to manipulate objects to present the barcode at the reader.

The price is particularly high for active tags because they have a battery.

4.5.2.2. *Physical perturbation*

RF tag reading is disturbed by the presence, for example, of metals in their immediate environment. Solutions must be considered case by case to minimize the disturbances, akin to what has been done for example with the identification of gas bottles.

4.5.2.3. *Mutual perturbations induced by labels*

In many applications, multiple RF tags can simultaneously be present (voluntarily or involuntarily) in the field of the reader. This can happen in a store at cash desks or between anti-theft gates. In the latter case, with technology, it is easy to detect a stolen object, i.e. a tag not inhibited by a cash desk. More complexity is the need to identify and read the contents of multiple tags in a field without missing any part. For this, readers use anti-collision algorithms or techniques.

4.5.2.4. *Sensibility*

In some circumstances, RFID reading systems are sensitive to interferences emitted by electromagnetic waves in computer equipments, e.g. computer screens or lighting systems and more generally by electrical equipments. The use of RFID tags must be tested taking into account a given environment.

4.5.2.5. *Unprintable*

With the barcode technology, it was possible to print the barcode of an item on an invoice or on the warranty of a product. With the RFID technology, this possibility is annihilated because the information is contained in the RFID and cannot be transmitted.

4.5.2.6. *Regulation*

There are now some questions about the impact of radio frequency on health. These questions have been debated for some years, especially as regards anti-theft gates and mobile phones. Passive tags are themselves safe, whatever their number, since they do not emit waves unless they are in the field of a reader. The studies focus essentially on the reader and aim at defining regulation criteria relating to avoidance of transmission power which create disturbances on equipments for health such as pacemakers, and also on the human body. A second important point to discuss regarding regulations is the respect for individual freedom. RFID technology is the subject of intense controversies. In particular in US, associations of consumer and liberty defence critize the technology since it could infringe individual freedom. They demand a moratorium on the deployment of this technology until legal safeguards are established. In France, the CNIL has taken a stand the subject. In a press release dated October 30th, 2003, Philippe Lemoine, Commissioner of the CNIL, co-chair of Galeries Lafayette and co-director Barcode EAN France, calls for vigilance: "While it is a major economic sector of retail, RFID is a potential threat to the privacy of individuals". At the international level, the companies, initially involved in the AutoIDcenter and grouped today in EPCglobal, make an important attention to the reserves expressed by consumers and relevant authorities concerning the privacy. EPCglobal has already provided a number of answers by encouraging businesses to report the presence of an RFID on a product or a packaging, to inform consumers about the possibility of neutralizing the tag and ensure the compliance of data stored on the network with all applicable laws regarding the protection of personal data.

4.6. Description of RFID applications

RFID tags are often considered an alternative way to replace and improve barcodes of the UPC/EAN standard. Identifiers are in fact long enough and uncountable to be able to give a unique number to each object, while currently used UPC codes can only give a number for a class of products. This property allows RFIDs to track the movement of objects from one place to another and from production chain to final consumer. Thanks to this property, many industrial companies of the supply chain consider the RFID technology as the ultimate technological solution to all problems of traceability, essential concept since emergence of sanitary crises related to food chains. Tags associated with objects will analyze themselves the environment around them, and will be able to communicate among them to make the management easier. The properties of tags also allow us to consider applications for the final consumer. Here is a non-limited list:

– refrigerator will be able to automatically recognize the products contained in it and check deadlines to optimize perishable food products use;

– sweater indicates its wash temperature in a washing machine;

– dish indicates its cooking time in a microwave;

– yoghurt continues to verify the cold chain;

– animal identification by implanting a tag (already mandatory in some countries, such as for dogs and cats in Belgium);

– identification of products for faster passage at cash desks in stores, especially in supermarkets;

– protection against abduction of newborn infants in maternity hospitals (e.g. the Montfermeil clinic (France) uses bracelets with RFID tags);

– in documents of an official nature, tags can be implanted to validate passports and visas, also for driver's licenses (for example to count the number of remaining points).

In the last part of this chapter, we will describe concrete application cases that have been implemented, some of them already in use.

4.7. Application examples

In this section, in order to illustrate the previous discussions, we will describe concrete examples of applications using RFID tags. For each example, the advantages and disadvantages of this technology will be listed. Note that all the mentioned applications are not in a "industrialization" phase: some of them are in a testing phase and others are only at a conception stage.

4.7.1. RFIDs in commerce

4.7.1.1. Consumer: Wal-Mart supermarkets

In order to introduce advantages of RFID tags in a supermarket, we will present a practical case: Wal-Mart [TUT 06]. This is one of the leading American companies. This position is partly due to the optimization of operating costs. As we will see, the use of an information system based on RFID tags aims at reducing costs. The goal is to improve the whole supply chain by controlling in an optimal way each component and by offering a possibility to the company to have a complete follow-up of sold products. The main contribution of RFID tags is on the Wal-Mart/providers relationship. Providers equip their ranges and/or all their products with tags identifying them. Once they arrive at the Wal-Mart storage center, the tags are immediately scanned and their IDs are stored. Very quickly, without unpacking the storage ranges, an inventory of received products is done. After the first step, the products on the shelves are scanned again: it is possible to update an inventory of products in shelves and in storage. Once a product is purchased by a customer, it is removed from the list of products on shelves. When a product runs out on shelves, the operators of the storage center are automatically alerted and are responsible for ordering them. The inventory management with the use of RFID tags does not stop here because Wal-Mart is constantly connected to its providers. They are able to know the status of stocks and their location. Therefore, using such a system allows Wal-Mart to reduce the number of errors and the costs for stock management. The company can adapt its offers to the requirement of its customers and be much more reactive to fluctuations in stocks. Inventories on shelves are also much faster and can have an overview of anomalies (thefts and out-of-date, obsolete, deteriorated products). In the future, other information could be added to RFID tags. For example, expiration dates allow them to quickly assess which products are unfit for consumption and to quickly eliminate them.

4.7.1.2. *Pharmaceutical industry: example of "CERP"*

CERP is a company providing drugs to many pharmacies. It has to manage a lot of stocks and products to deliver each day. To appreciate the performance improvements given by the RFID technology, the company has decided to equip a testing storage center. There, each box of drug is equipped with an RFID tag. These tags are read once the box is prepared. They are identified again at shipping and delivery stages. This method allows them to keep an updated inventory of the stock status: this is particularly important for drugs produced at very low quantity, rarely ordered and quickly outdated. All this information is easily retrievable, and at a minimal cost. Such an architecture in CERP also provides the possibility to perform accurate statistics in order to manage the drug stock more efficiently.

4.7.1.3. *Other applications of RFID in commerce*

In general, commerce gains a lot with the use of this technology: it can improve performances at all levels in the supply chain. It becomes possible to significantly reduce the risk of human errors and costs in a general case, while offering customers new services and improving their satisfaction. Many other applications are already deployed or are in a prototype stage, while others are still at a conception level in a commercial context [COM 06], for example:

– widespread use of RFID tags as an anti-theft system (already deployed in big companies);

– a clothing store could display information in a fitting room and advice on products that a client tries out;

– a vendor can, with the help of a reader associated with a hand-held computer, get a multitude of information about a product, check its availability in stocks and perform orders;

– RFID tags can be affixed on products to track their history, their location and manage stocks;

– automatic cash desk: client passes products facing the cash desk and pay. He does not need an operator to pay for purchases;

– after-sales services can track products equipped with RFID tags: a company can be provided with a history of a tag if it is authorized to do so (it must be authenticated).

In a more or less near future, retail and commercial fields will be transformed by the ubiquitousness of RFID tags. However, several facts tarnish this positive result:

– anti-RFID associations advocate for the abolition of this technology;

– the protection of privacy must be ensured (an attacker should not be able to track a person by identifying the carried tags). This problem can be solved by the use of cryptography;

– the use of this technology is most effective when several participants in the supply chain have adapted their working methods;

– large investments must be made to integrate RFIDs into an already existing information system;

– the price of an RFID tag is clearly higher than that of a barcode. A massive deployment on all products is currently unrealistic (for example, today it is not economically viable to equip all milk cartons with RFID tags).

4.7.2. *Access control*

Access control consists in checking whether a person or entity wanting to access a resource has the right to do it. The RFID technology is increasingly used in this field.

4.7.2.1. *Access to a site*

In the case of a sensitive company or site, access management is critical. There are solutions that take advantage of new technologies: a person with access authorization has an RFID tag allowing him to enter. A reader checks the rights of the person by reading the tag and opens the gate if they are sufficient.

4.7.2.2. *Transports*

Control in transport: the Navigo pass of the transport network in Paris allows users to quickly pass through security gates. The latter ones verify information of a card holder, authorized travel zones, date of expiry and grant access to the transport network if the rights are sufficient. This increasingly widespread system is used in other cities both in France and other parts of the world.

4.7.2.3. *Events*

In event management (concerts, sporting events, etc.) recently, several ticket counterfeiting cases have occurred. Tickets (mostly printed on paper) are easily forged and the authenticity verification takes time, as the use of RFID tags in such a situation allows us to quickly check tickets and limit the risk of creating false tickets. This particular solution was chosen for the football World Cup in 2006 and the Olympics in Beijing in 2008.

4.7.3. *Culture and RFID*

4.7.3.1. *Libraries*

In the case of a library, equipping books with tags greatly facilitates the management of inputs and outputs. The applications are numerous:

– inventories are easily and quickly done;

– all CDs/DVDs are equipped with tags and it becomes possible to verify the contents of a box without opening it;

– an anti-theft feature can be added to tags;

– lending and returning procedures become easier. It is also possible to implement automatic return terminals.

Today, several libraries in Paris use the RFID technology in their information systems to facilitate the management of paper and digital works. Despite the advantages, several negatives points have been reported:

– some products contain metallic elements that perturb the proper operation of tags. This is particularly the case with some digital books and products such as CDs/DVDs;

– initial investment is relatively significant: tags must be installed on all works proposed by the library;

– the anti-theft feature of RFID tags has limited effectiveness: in some cases, it is not difficult to remove a tag from its support;

– again, the price of such an installation can be a deterrent;

– libraries in Paris do not use the same RFID technology and it is not easy to ensure the interoperability (a person can borrow and return books in different libraries).

4.7.3.2. *Other applications*

RFID can also be used for informational purposes. For example, some art galleries equip their pictures with tags readable by a Bluetooth pen connected to a PDA. A visitor can make his visit as he wishes while accessing information on the exhibited works. Museums can also use similar techniques: a visitor is given a box containing an RFID reader upon arrival. When he is front of a work, the box starts to diffuse corresponding information, for example under the form of sound (with headphones) and images (on a PDA). This procedure allows them to offer more live and interactive tours without requiring a guide. The services offered are thereby very useful.

4.7.4. *Payment*

In this field, applications using RFID technology are currently less. Presumably, this lack of enthusiasm is due to a lack of confidence of various participants. Systems that minimize risks are being studied to reassure unresponsive companies.

4.7.4.1. *Services for drivers*

In Singapore, some roads are subject to a tax: drivers carry an RFID tag identifying them and their account is automatically debited when they use a toll road (bounded by toll booths). A similar system is used by some gas stations: a driver has to present only his ID and fill fuel. It is then automatically billed. In both previous cases, it is important to reduce as much as possible, the operating range of the reader: the driver must be charged, not any other person bearing a tag and walking near the reader. We can also cite a service of automated toll payment which was implemented in France: "Liber-t". This service allows one to have a greater fluidity on all toll roads in France. This service also provides simplicity, comfort and serenity to drivers: it is no longer required to pull down the car window, take a ticket, pay by card or search coins for change.

4.7.4.2. *Contactless payment*

The rare contactless payment systems are limited to small amounts. CROUS (a student social support service in France) has deployed in its restaurants at universities a solution based on contacless Moneo [MON]. Other systems merely identify the tag of a person and then charge him (by linking the identity to a customer number). Several similar systems are in development or in testing phase (see [L'A 08]). However, one factor that slows down the expansion of contactless payment is the need of a support system on a large scale: what is the interest of a consumer if he can not use his contactless card in the nearby shops? To be used and to survive, the technology needs to be massively deployed in shops.

4.7.4.3. *Case of nightclubs*

Several nightclubs in the world (particularly in Netherlands and Spain) equip their VIP customers with RFID tags. They have a special VIP area where they can enter and pay through their RFID tag (most of the time, the tag has a size of a rice grain and is injected under the skin). Therfore, it is very easy for them to order drinks: it is no longer necessary to carry a payment card or any document certifying the membership to the VIP club.

4.7.4.4. *Application in hotels*

Using RFID technology, a hotel can suggest new services to its customers and greatly simplify the "entry" procedure: a client no longer has to carry keys, tickets for a meal, a consumption or a breakfast, or even card to enter the hotel. He will merely present the RFID tag to enter his room and access services.

4.7.5. *RFID and health*

The medical world is a field where benefits from the RFID technology are enormous. Many repetitive tasks of various participants in this field (e.g. nurses) can be automated, which will significantly reduce human errors.

4.7.5.1. *Anesthesia dosage*

During surgery, the anesthetist doctor is responsible for administering an adequate anesthetic dose. This is a sensitive operation since the dose should be sufficient in order that the patient is put to sleep correctly and does not move, but it should not be too strong, otherwise it could cause irreversible damage and/or troublesome side effects.

To assist the doctor in the amount to be used for a patient, a company proposes to intravenously administer an anesthetic. The doctor must strive to maintain the same amount of product in the patient's blood. For this, several syringes are available, each contains a different dose. The use of RFID tags can automatically determine the dose to be injected based on the information on the state of the patient. Obviously, to have a real interest, these tags must be correctly integrated into the information system. The system forbids the injection with a syringe other than indicated by calculation. This application aims at reducing the possibility of human error(s) and giving the patient an adjusted dose, while reducing side effects.

4.7.5.2. *Management of blood samplings*

Related to blood donations, samplings for testing or management of samples, the work is important, tedious and prone to many errors. The use of RFID tags can facilitate the management process.

Blood samples can be equipped with tags. An RFID reader can find information about blood (essentially the blood group and rhesus). So when Mr. Dupont has to receive blood from Mr. Dupuis, the operator uses a computer equipped with an RFID reader to verify the correct matching of blood for transfusion. In general, this approach significantly reduces human errors. Note that the use of computer allows us to hide the identity of the person giving his blood: only the computer is able to know his identity. Although barcode-based systems are currently used, it does not prevent errors: the reduction of hospital staff forces employees to work faster and more efficiently. It is also important to note that in the case of a major crisis (i.e. at a moment when many blood transfusions are needed), the use of RFID is very effective: with less manual controls, a staff can focus his attention on procurement, and gains in efficiency.

Finally, the use of RFID tags in this field allows a quality check at any time during the lifetime of a sample. Again, one can generate accurate statistical data with the use of this system.

4.7.5.3. *Drug tracking*

Today, there are more and more cases of counterfeit drugs, which are potentially dangerous. The use of RFID technology can accurately track and verify "the identity" of a drug. The use of barcodes, as we are doing now, is not reliable: this technology is easily falsifiable. In addition, pulling out the tag of a drug is not a solution: no tag, it must be considered potentially dangerous.

Obviously, in order for the protection to be effective, each drug box and plate should be equipped with a tag. Each tag contains a unique identification code, a serial number, an expiration date, a production date or any other information to verify the integrity of a drug. In this way, each participant in the supply chain will ensure the compliance of a drug. By consequence, fraud, expired drugs and placebos can be avoided. However, this technology has (as always) a cost: poor countries are the most affected by drug trafficking and are now unable to equip themselves in the use of RFID.

4.7.5.4. *Food tracking*

Due to recent threats related to food (mad cow disease, bird flu, etc.), it is increasingly important to be able to ensure the origin of a product. As we have previously seen, there is a possibility offered by the RFID technology, which aims at reassuring consumers: is the product labelled "bio" really "bio"? With RFID, it is possible to reliably track the origin of a product.

4.7.6. *European biometric passport*

The new European biometric passport uses RFID technology to enhance the security of identity papers. Information, such as citizen's name, fingerprints and a photograph, are stored in a tag. With the help of a reader, it is possible to authenticate a passport and access the information of the person who possesses it.

4.7.7. *Future perspectives*

Most of the examples cited above and the current applications are only cases of integration in small scales. Developed and implemented architectures are adapted to a number of specific needs. The future of RFID is an interconnection of all tags in the world, giving access to a huge range of possibilities which are still unexplored. We can essentially think about the following benefits:

– a better management of customers/suppliers data and stocks with a common technology to all (at a national and/or global level);

– a standardized and global architecture will give an opportunity for different users of RFID technology to easily interact with each other;

– better tracking of products during their lifecycle.

Generally, such an architecture would greatly simplify exchanges between entities. The standardization of such a technology would enable all to exploit better utilize their benefits. Finally, note that the list of applications using RFID tags is only limited by imagination.

4.8. Conclusion

According to the points presented throughout this chapter on applications two opposing theories can be released in the coming years about the development and implementation of this technology. At a first step, we can consider a negative future and thus a stagnation or even a development discontinuation of the RFID technology. On one hand, supermarkets do not invest in RFID readers and tags because they are very costly. On the other hand, organizations like CASPIAN (Consumers Against Supermarket Privacy Invasion and Numbering) [CAS] can prove that RFID tags are dangerous for the privacy of consumers. Finally, scientific studies show that radio waves can be harmful to health. Thus, RFID tags are being boycotted. Producers of these technologies are unable to sell their goods and therefore cannot invest in research and development. The RFID project is abandoned. RFID tags cannot be developed and are neglected. Barcodes remain the leader of the market.

In a second step, we can consider a very positive future, resulting in a strong development of the RFID technology. Regarding publications, both in terms of articles and patents, this technology is booming. The application fields of RFID are widened (access control of persons by identification and authentication, contactless payment etc.), which reduce production costs and therefore the price of tags. The distribution centers are now equipped with RFID tag readers.

But apart from these two scenarios, the evolution of some problems or issues will tip the balance toward one of the theories presented above. For this, we should consider works and laws concerning the privacy of consumers and the effect of radio waves on the human body. As we have seen previously, the RFID technology could be dangerous for individuals. For example:

– possibility of privacy invasion in case of eavesdroppers;

– using information contained on tags of passports to select persons of some nationalities;

– recording information of citizens according to their consumption habits in whatever domain (food, culture, etc.);

– potential problems of digital/economic sovereignty related to the infrastructure of the EPCglobal network;

– the case of chips under skin naturally raises ethical questions and reports about the right to physical integrity. The limitation to voluntary does not ensure a sufficient guarantee, any person who refuses such subcutaneous tags has a risk of being a victim of discrimination;

– identifying a person by a signature of all his RF tags (bank cards, mobile phone, travel card, etc.);

– generation of RF signals can be dangerous to health or can interfere with the operation of biomedical devices or act on receptors in human body.

4.9. Bibliography

[CAS] CASPIAN, http://www.nocards.org.

[COM 06] COMMEAU C., http://www.journaldunet.com/solutions/0603/diaporama/magasins-futur/1.shtml, 2006.

[EPC] GS1, http://www.epcglobalinc.org.

[GOM] GOMARO, http://www.gomaro.ch, 2010.

[ISO] International Organization for Standardization, http://www.iso.org.

[L'A 08] L'ATELIER-BNP-PARIBAS, http://asie.atelier.fr/telecom/mobilite/article/sur-le-vif-une-carte-de-credit-en-porte-cles, 2008.

[MON] Moneo, http://www.moneo.net.

[TUT 06] Walmart and RFID: A Case Study, http://www.tutorial-reports.com/wireless/rfid/walmart/case-study. php, 2006.

PART THREE

Cryptography of RFID

Chapter 5

Cryptography and RFID

5.1. Introduction

To identify an RFID tag (generally a communicating contactless object, known as CLD for Contact less Device), multiple protocols have been proposed in recent years. These protocols must satisfy a number of security principles. In particular, a malicious person must not be able to usurp a legitimate RFID tag. Such an adversary may use different stratagems to be authenticated: listen to remote communications, interrogate tags, contact a reader, replay or use data obtained by these expedients, alter communications, use several tags during the course of the protocol, etc.

These properties can guarantee the authenticity of an RFID tag, but we must take into account the nature of wireless communications. Indeed, as it is possible to interrogate and listen remotely to communications emitted by RFID tags, the obtained data must not provide information about the person who has them. Without specific protections, it may be possible to track an RFID tag remotely and therefore, for example, to track the movements of individuals. To avoid such problems, it is necessary to introduce additional criteria in order to respect the privacy of individuals.

Today, the RFID technology is already deployed on a large scale. Some existing solutions are based on proprietary solutions, i.e. they use algorithms with specifications which have not been published. Some of these proprietary solutions have hit the headlines recently because they have experienced setbacks.

Chapter written by Julien BRINGER, Hervé CHABANNE, Thomas ICART and Thanh-Ha LE.

For example, the DST (Digital Signature Transponder) system allows the control of the car door opening and gas purchase payments. There were more than 7 million manufactured DST tags, which were usable in 10,000 gas stations. In USENIX Security Symposium 2005, [BON 05] it was explained how to break this system. The attack started with a reverse engineering of the component, and the size of the implemented key were then revealed to be too short.

More recently, history has been repeated with the MIFARE Classic 1k. The present cryptography has enabled us to provide mutual authentication between a reader and tags but also the data access control on the card. The MIFARE system has been mainly used for the payment in the public transport in London and Amsterdam. A proprietary encryption algorithm CRYPTO1 was used. Various researchers, with different approaches, were able to find out step-by-step the details of the algorithm CRYPTO1 [GAR 08, NOH 07, KON 08]. Finally, in ESORICS 2008, [GAR 08] gave a complete description of the algorithm CRYPTO1 which revealed its inadequacy in terms of its cryptographic strength.

These examples (see also section 5.4.3) motivate us to follow a more academic approach where the protocols are based on more solid published bases [KER], as we will do in this chapter. Thus, we describe first a security model for identification protocols. This description also takes into account the notion of privacy. Then we give examples of different protocols proposed in the literature. We conclude with a brief overview about physical attacks of RFID tags.

5.2. Identification protocols and security models

In this section we firstly define and introduce the notations used to describe the identification protocols. We then present in a formal way the security notions. Finally we will detail the notions of privacy and the problems that they formalize.

5.2.1. *Definition of an identification protocol*

An identification protocol is a mechanism between two entities using predetermined algorithms, denoted by $\mathcal{P}$ and $\mathcal{V}$. The algorithm $\mathcal{P}$, called Prover, aims at proving the algorithm $\mathcal{V}$ has a numerical identity. The algorithm $\mathcal{V}$, the Verifier, aims at verifying the coherence of this proof. The identity of $\mathcal{P}$ is a numerical value known at least by $\mathcal{V}$.

A protocol between $(\mathcal{P}, \mathcal{V})$ is an exchange of messages. The last message is sent by $\mathcal{P}$ in the case of an identification protocol to allow $\mathcal{V}$ to finish the identity verification. In general, the execution of a protocol requires only two or three message deliveries. If necessary, the protocol can be run consecutively a multiple number of times to ensure the authenticity of $\mathcal{P}$ accurately. We term the transcript of a protocol, as messages exchanged during an instance of the protocol.

According to the cryptography types, Prover and Verifier possess and share different information and identities. In the symmetric case, each Prover $\mathcal{P}$ possesses a unique and secret key $K_{\mathcal{P}}$, which identifies it uniquely, and the Verifier also knows this value (for all provers of the system) to verify the proof performed by the Prover. In the asymmetric case, $\mathcal{P}$ possesses a public/secret key pair $(K_{\mathcal{P}}^p, K_{\mathcal{P}}^s)$, and the public key corresponds to its identity. In this context, $\mathcal{V}$ knows only the public key and the protocol allows $\mathcal{P}$ to prove that it knows a secret value related to its identity $K_{\mathcal{P}}^p$.

In both cases, there exist a VERIFY function associated to the protocol, which enables $\mathcal{V}$ to verify the authenticity of a transcript. The function is applied on a protocol transcript, an identifier of $\mathcal{P}$ and probably other values such as a secret value related to $\mathcal{V}$. This function returns a bit that indicates if the transcript is authentic. This function can be used to identify $\mathcal{P}$. In this case, $\mathcal{V}$ uses VERIFY on the transcript and for all the different identifiers ($K_{\mathcal{P}}$ or $K_{\mathcal{P}}^p$ depending on the case). This method is not optimal in terms of computing time. In some cases, there exists a COMPUTEID function which makes it possible, from the transcript and other values related to $\mathcal{V}$, to find the identifier of $\mathcal{P}$.

5.2.2. *Classical notions of security*

5.2.2.1. *Correctness*

The correctness of a protocol ensures that a honest prover is identified with a probability very close to 1, i.e. the protocol is an identification protocol. To establish it, it is usually sufficient to formally verify the correctness of the calculation made by VERIFY or COMPUTEID.

5.2.2.2. *Impersonation resistance*

It is also necessary to ensure that an illegitimate prover cannot be accepted. Hereafter, we call this illegitimate proof an adversary and we denote $\mathcal{A}$ an individual trying to find security holes in the identification protocol. The protocol is said to be attack resistant if an adversary cannot produce a valid transcript, i.e. accepted by VERIFY, which is not identical to a transcript previously observed. Different types of adversaries exist to represent different levels of threats. It is particularly interesting to define their different modes of action:

– passive adversary: knows only the public information and cannot interact with $\mathcal{P}$. Therefore he is very limited in his possibilities;

– eavesdropper adversary: can interrogate a prover and obtain information only by listening to communications between $\mathcal{P}$ and $\mathcal{V}$;

– active adversary: can communicate with the prover either in an isolate manner or by using them to respond to messages sent by the verifier to obtain information. This latter type of attack is a *Man In The Middle* (MTM) attack;

– concurrent adversary: they are active adversaries, who can create many protocol sessions with the same prover at the same time.

These different possibilities are related to real threats. For example, for some types of tags, a passive eavesdrop is possible upto until several meters. Passive or eavesdropper adversaries represent adversaries who do not have RFID tags but possess the technology to simulate and listen to them. Active or concurrent adversaries have a large number of RFID tags and can use them to attack the security protocol.

NOTE.– *Our model does not take into account relay attacks where a dialog between a legitimate tag and its reader is retransmitted [HAN 05]. Similarly, network attacks such as denial of services or viral contamination from the RFID system [RIE 06] are not considered here. Nevertheless we signal the protection of RFID guardian [RIE 05], firewall for an RFID system which determines the presence of external emissions in the band of wavelengths of tags.*

5.2.2.3. *Zero-knowledge protocols*

An identification protocol based on asymmetric cryptography is often a proof of knowledge of the private key associated with its identity. It is then important that this evidence does not reveal information about the private key. Such a proof is said to be zero-knowledge (ZK) if it is possible to demonstrate that the verifier has no information on the private key of the prover. In the case where such proof is an interactive protocol, we talk about ZK protocol.

To demonstrate that this property has been achieved, we can use a prover simulation technique. It is necessary to establish the possibility to simulate the behavior of a prover without knowledge of its private key. A ZK proof can be to honest or malicious verifiers according to the type of verifier face to the simulator. If he is honest, the verifier chooses messages regardless of the messages sent by the prover. Otherwise, the verifier adapts its choice strategy in function of messages sent by the prover to be able to learn information about the secret.

The study of ZK schemes is very important in cryptography and we will give examples of the schemes which can be used for RFID.

5.2.3. *Privacy notions*

Privacy is a very important notion in the context of cryptographic schemes associated with RFID. A protocol respects privacy if it is possible to prove that the holder of an RFID tag is anonymous and is not traceable by other persons.

5.2.3.1. *Different privacy leakages*

The first type of privacy leakage is the ability to retrieve information related to a person. This problem occurred for example on early versions of the Belgian electronic passport [AVO 07]. More generally, the protocol is not anonymous if it is possible to find information related to the tag. For example, if it is possible to find its identity through simple eavesdropping or by active attacks, then there is a privacy leakage.

There are other possibilities of privacy leakage. For example, even if all values associated with the tag cannot be calculated, an adversary should not be able to determine if multiple transcripts of the protocol have been issued by the same RFID tag. If possible, then an adversary can easily follow a remote person. Many other scenarios are possible. If an adversary is able to corrupt a tag and get its internal state associated to the secrets, then he can try either to determine the source of transcripts issued in the past or to follow the future exchange of the RFID tag.

If it is impossible to determine old transcripts, then the scheme is called *Forward* privacy. If it is not possible to determine the source of new messages, it is called *Backward* privacy. In addition, this adversary can use the result of the function VERIFY to obtain information. Indeed, after the modification on the fly of the communications emitted by a tag, the positive or negative results of the evaluation of this function can give a lot of information such as in the case of the protocol presented in section 5.3.1.6.

5.2.3.2. *Privacy definition*

Among various formal models on this topic, we summarize here the model proposed in ASIACRYPT 2007 [VAU 07]. The main idea of this model is to analyze the privacy as a possibility to simulate a system. Indeed, if an adversary is unable to differentiate the real system of a system that is simulated without any knowledge of secret information of the real system, then it implies that there is no information leakage.

NOTE.– *[JUE 07, LE 07] propose different privacy models. The one we have chosen seems more general.*

The model takes into account different possibilities of privacy leakage introduced in the previous section. They depend on the actions authorized to an adversary. These actions are represented through oracles that the adversary can interact with.

Oracles. The adversary can effect calls to the following oracles:

– CREATECLD(SN) creates a new RFID tag associated with a serial number (SN);

– DRAWCLD allows a probabilistic selection of a chosen number of tags among the ones created by the previous oracle and allocate randomly a virtual Serial Number (vSN) to them. The correspondence between the vSN and the SN is kept secret in a table $\mathcal{T}$. The SN enable the identification of a tag in the system while an adversary sees only the vSN. There is, for example, a threat if the adversary can link different vSN to the same SN;

– FREE(vSN) liberates the used RFID tags by DRAWCLD;

– SENDVERIfiER (respectively SENDCLD) allows an adversary to interrogate a verifier (resp. an RFID tag) by providing a message as input;

– VERIFY verifies whether a transcript corresponds to a normal execution of the protocol enabling the identification of a tag;

– CORRUPT returns the internal memory of an RFID tag.

Adversaries. According to these opportunities, there are different types of adversary:

– a *Strong* adversary has access to all oracles. In particular it can corrupt an RFID tag with CORRUPT to obtain the memory and then reuse the attacked tags to follow them;

– a *Forward* adversary may use one and at only one time the oracle CORRUPT. Once this call is performed, the system is considered destroyed, it has no longer the right to use other oracles and can use only information of the memory of the target tag and the transcripts that he has previously acquired;

– a *Weak* adversary is not allowed to use CORRUPT;

– a *Narrow* adversary is not allowed to use VERIFY.

The narrow adversary can be combined with the strong, the forward and the weak adversaries, so that it is possible to define six types of adversaries and six forms of privacy. For example, a scheme is said to be *Narrow-Forward* private if no *Narrow-Forward* adversary can attack the privacy on the scheme.

Concretely, the privacy is defined as the impossibility of an adversary to distinguish a real RFID system from a simulated system. To verify this, an algorithm called a simulator and denoted by $\mathcal{B}$ (for blinder) is defined. This algorithm must simulate the SENDVERIfiER, SENDCLD and VERIFY oracles and without control on the function DRAWCLD. This requires that $\mathcal{B}$ does not have advantage over SN RFID tags. In particular, it cannot link a numerical identity to an RFID tag because it would link it to a vSN which may change the corresponding tag without knowing it. To complete this definition, we must define a set of attacks (where the probability of success of an adversary is measured): it is a theoretical representation of different possible actions of the adversary to try to guess which is the system with which it interacts. In this context, the adversary is connected to the real system, then facing the simulated system. There is a lack of privacy if the adversary can distinguish the two phases. Therefore, a protocol respects privacy if we can prove that these two phases are indistinguishable. In particular, a distinction would be made on the SENDVERIfiER, SENDCLD and VERIFY oracles simulated by $\mathcal{B}$.

It should be noted that calls to the CORRUPT oracle are not simulated by $\mathcal{B}$. Indeed, after each DRAWCLD, if the adversary uses CORRUPT in the case of a real system, it obtains the internal state of RFID tags and thus can continue to track after each call to CORRUPT. Whereas in the case of a simulated system, as $\mathcal{B}$ has no advantage on DRAWCLD, it cannot simulate them several times in the same way. The model examines only the influence of other oracles on the privacy.

A very useful lemma was established by Vaudenay in [VAU 07]. It links safety and privacy.

LEMMA.– *A secure scheme against impersonation attacks and narrow-weak (resp. narrow-forward) private is weak (resp. forward) private.*

This result is based on the fact that if the protocol is secure, then the use of the function VERIFY provides no information because it is impossible for an adversary to generate successfully the messages that will be accepted by this function.

5.2.3.2.1. New types of adversary: with public identity and with hidden identity

Recently, [BRI 09] has introduced two new types of adversary related to identification protocols with public key. Indeed, the list of public key of RFID tags can be either freely distributed or kept secret. An adversary with access to this list can use it to follow a tag with Public Identity (PI). If it is not possible, it will be called Hidden Identity (HI).

NOTE.– *In the case of a strong attacker, this distinction is not useful. Indeed a strong adversary knows all the secrets and can therefore calculate all public keys.*

5.3. Identification protocols

We describe afterwards several examples of protocols such as symmetric protocols and asymmetric protocols which ensure different levels of security and privacy. In addition to the notions of security, efficiency is also an important factor. It particularly concerns the extensibility problem (or scaling) for a system containing a large number of RFID tags. For N tags, the protocol is considered scalable if a player does only $\mathcal{O}(log(N))$ operations to identify a tag. We denote by L the list of identities of tags contained in the system, this list in is usually stored in the verifier.

The cost of implementation in tags is also a problem to be taken into account. For the cost, we must consider two main implementation constraints. The first constraint is with physical space. To compare the space requirements, we often use the notion *Gate Equivalents* (GE). A GE is equal to the space of an NAND gate with 2 inputs. The second constraint of a RFID tag is the power consumption of the component. Depending on the transmission mode, the power consumption can be restricted to just a few μA.

5.3.1. *Symmetric cryptography-based protocols*

This kind of scheme is widely studied because symmetric cryptography ensures very efficient hardware implementations. In this context, central system and tags share a certain number of keys and a tag must prove the knowledge of identifier keys. Until now, only one of these schemes (see section 5.3.1.5) allows privacy. These symmetric protocols cannot, in general, achieve the expected level. By consequence, the reader must make a number of calculations proportional to the number of RFID tags in the system. There are some exceptions, notably the work of Molnar *et al.* [MOL 04]. However this scheme has a specific feature: keys are shared between tags. Therefore, the corruption of a tag provides information of keys in other tags.

5.3.1.1. *Cryptographic properties*

In the protocols presented below, we will use hash functions. A cryptographic hash function [MEN 96] is a transformation taking as input a message of arbitrary length and returns a value of relatively small fixed length, called the hash (or condensate) of the message.

Let $H: \{0, 1\}^* \rightarrow \{0, 1\}^n$ be the function accepting any input binary message and returns a hash of l bits. This function must have the following properties:

Pre-image resistance. The pre-image resistance (or *Universal One Wayness*) ensures the impossibility of finding an antecedent (or pre-image) given a hash: given a value y in $\{0, 1\}^n$ of pre-image unknown by the adversary, the latter cannot, except with a negligible probability, find or calculate a message m' such that $H(m') = y$.

This should in particular be true for $y = H(m)$ if m is unknown by the adversary.

Second pre-image resistance. This property refers to the difficulty of finding a second message corresponding to the same hash of a known message given a randomized message m, an adversary cannot find, except with a negligible probability, a message m' different from m such that $H(m) = H(m')$.

Collision resistance. This property extends the previous one in order to provide such a situation regardless of the choice of m: an adversary cannot find, except with a negligible probability, m and m' such that $H(m) = H(m')$.

5.3.1.2. *Implementation cost of symmetric primitives*

5.3.1.2.1. Hash function

The SHA-1 algorithm [NAT 95], while waiting for the definition of its successor at the end of the SHA-3 [NAT 08] competition, is used as one of the main hash functions because it was defined by a standard of NIST. Therefore, it is natural to study its possible implementation for the RFID. SHA-1 is designed for 32 bits platforms. However, there are some works that study architectures with limited resources. [O'N 08] defines an 8-bit design by transforming all 32-bits operations to 8-bit operations. The implementation requires then 5527 GE and consumes 2.32 μW at 100 kHz. It is also possible to imagine a specific hash function to be implemented in RFID tags, for example the case of [SHA 08].

5.3.1.2.2. Encryption algorithms

The symmetric cryptographic protocols that we present in the next section do not use encryption algorithms, they use hash functions. However, encryption algorithms may also be used according to particular modes [MEN 96] which allow them to have certain properties of hash functions. In addition, it could be interesting to use them to ensure data confidentiality. AES (*Advanced Encryption Standard*) is the cipher block adopted as an encryption standard by the NIST [NAT 01] since 2001 to replace DES. Proposed by Daemen and Rijmen, it became one of the most popular symmetric algorithms. Although it is used in many applications, hardware implementations of the algorithm focus mainly on the throughput optimization. However, in 2004, Feldhofer *et al.* [FEL 04] presented an adapted implementation to the RFID environment. The result is encouraging. On a 0.35 μm CMOS technology, the design has a power consumption of 8.15 μA and the complexity in hardware is estimated at 3595 GE [FEL 04]. A new version of this implementation has been published [MAR 05] where measurements on a real component show that 3400 GE is necessary and its power consumption varies around 3 μA.

PRESENT [BOG 07c] was designed according to the principles that led to the creation of AES but with RFID tags the target implementation. In the original version, PRESENT-80 (80 bits key size) is implemented on a surface of 1570GE and PRESENT GE-128 is implemented on an area equivalent to 1886GE. Recently in 2008, a new implementation of PRESENT [ROL 08] based on a serialized architecture was presented. Thanks to this modification, the surface can be reduced to 1000GE.

Other examples of functions and implementations of RFID have been proposed as HIGHT [HON 06] (3048GE), mCrypton [LIM 05] (3500-4100 GE for encryption and decryption and only 2400-3000 GE for the encryption) and DESL [LEA 07] (1848 GE).

The European project eSTREAM has studied the safety and effectiveness of various stream cipher algorithms to define a portfolio of recommended algorithms during 2008. The list includes three algorithms for a hardware implementation on a component with limited resources: Grain, MICKEY and Trivium. [FEL 07b] made a detailed comparison between Trivium and Grain in terms of implementation. He proves that for a 0.35 μm technology CMOS, at the same level of security, the average power consumption of Grain is 0.68 μA, 0.80 μA for Trivium and physical space is 3360 GE for Grain and 3090 GE for Trivium (see also [FEL 07a]). Other studies have been conducted to assess the various stream cipher functions proposed during the eSTREAM project and are available on the project's website: http://www.ecrypt.eu.org/stream.

To summarize the different elements, a summary of characteristics mentioned in these implementations is presented in Table 5.1.

Primitives	Space (GE)	Power (μW)	Average current (μA)
SHA-1 [O'N 08]	5527	2.32	?
AES [MAR 05]	3400	?	3
PRESENT [ROL 08]	1000	11.20	?
Grain [FEL 07a]	3360	?	0.68
Trivium [FEL 07a]	3090	?	0.80
HIGHT [HON 06]	3048	?	?
mCrypton [LIM 05]	3500-4100	?	?
DESL [LEA 07]	1848	8.0	0.89

Table 5.1. *Some implementations of symmetric primitives*

5.3.1.3. *WSRE*

Introduced by Weis *et al.* [WEI 03] and known as the *Randomized Hash Lock protocol*, a prover uses a hash function H to prove his knowledge of his secret key. The scheme is summarized in Figure 5.1.

Device$T$ Reader

Key: K^T The List of Key: L

$\xleftarrow{\quad a_0 \quad}$ pick a_0

pick r_1
$a_1 = H(a_0, r_1, K^T)$

$\xrightarrow{\quad a_1, r_1 \quad}$

Determine whether there exists in L
one key K verifying the equations
$a_1 = H(a_0, r_1, K)$

Figure 5.1. *The WSRE Identification protocol*

If H is pre-image resistant, then the scheme is secure against active impersonation attacks and is weak private.

However, the scheme is not *Narrow-Forward private* because the knowledge of the key $K^{\mathcal{T}}$ allows us to test directly whether a transcript (a_0, a_1, r_1) previously observed is linked to the corrupted tag by checking whether $a_1 = H(a_0, r_1, K^{\mathcal{T}})$. We also note a significant drawback: the scheme is not scalable since the verifier must check the final equality for all identities.

5.3.1.4. *MSW*

The scheme of Molnar *et al.* [MOL 05] has been proposed to avoid such inconvenience. It is scalable and in addition, allows us to delegate a part of verification rights to other verifiers. The scheme is described in Figure 5.3 and uses a specific architecture for the keys, according to the hash tree principle.

During system initialization, a Trusted Center generates a tree of secrets (keys), for example a binary one. Each leaf is associated to an RFID tag and each tag knows all keys $K_1, \ldots, K_d$ along the path from the root to its leaf. When a tag is interrogated by a reader by sending to it a random value a_0, the prover generates a new value $- H(a_0, r_1, K_1^{\mathcal{T}}), H(a_0, r_2, K_2^{\mathcal{T}}), \ldots, H(a_0, r_d, K_d^{\mathcal{T}})$ – where $r_1, \ldots, r_d$ are random values generated and transmitted by the prover. The Trusted Center can easily check which leaf corresponds to the received value in its tree of secrets by verifying as a given $(a_0, r_1, \ldots, r_d)$:

1) to which node corresponds $H(a_0, r_1, K_1^{\mathcal{T}})$,

2) which of the two children in this node is associated with $H(a_0, r_2, K_2^{\mathcal{T}})$,

3) then repeating this principle, level after level from the root to the leaves,

4) and then identifying which leaf (tag) is associated with $H(a_0, r_d, K_d^{\mathcal{T}})$.

Figure 5.2 illustrates this principle.

A concrete example is given in [MOL 05]: 2^{20} tags, a secret tree with a branching factor $Q = 2^{10}$ and is made of two levels. Each tag stores two 64-bit keys. Using a Tree-Based identification enables a verification in only 2×2^{10} hash computations instead of 2^{20} without the tree. More generally, if the system size is N, the number of computations is $O(log_Q(N)Q)$.

If H is pre-image resistant and collision resistant then the scheme is secure against passive impersonation attacks and *Narrow weak private*.

Contrary to the previous scheme, the security is only passive. The value a_i depends only on i-th round, hence it is possible to mix several transcripts sharing the same first keys to produce an altered transcript but still valid. In addition, the same technique allows us to test, via VERIFY, if provers have common keys, and therefore represent an violation of privacy (*Narrow-weak* adversary).

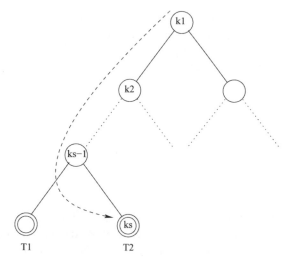

Figure 5.2. *One descent in a binary tree of secrets*

To correct this problem, a solution is proposed in [BRI 08b] to the different rounds by replacing the calculations $a_i = H(a_0, r_i, K_i^{\mathcal{T}})$ by $a_i = H(a_{i-1}, r_i, K_i^{\mathcal{T}})$. This leads, with the same assumption on H, to a secure scheme against active impersonation attack and weak private.

Nevertheless, a significant risk remains. If an adversary corrupts a tag then he gets all the keys of the path leading to the leaf associated with the tag. Thus the scheme is not forward private because the adversary gets private keys that are common to other tags and can partially track these tags. So the attack here is greater than for the WRSE scheme where only corrupt label is concerned. This threat has motivated the introduction of an additional protection by using POK (*Physically Obfuscated Key*, see section 5.3.3.1).

5.3.1.5. *Scheme of Ohkubo, Suzuki, Kinoshita*

The main purpose of this scheme [OHK 05] is *Forward private*. The main idea of the scheme is to modify the key at each identification. Hence, getting the key does not enable the adversary to track the past communications of a prover. The scheme is described in Figure 5.4. If H is collision and second pre-image resistant, the scheme is then secure against active impersonation attack and *Forward private*.

This scheme is not *Narrow-strong private* since an adversary getting the key of the prover is able to follow its future communications.

Efficiency is also at a disadvantage in this scheme because the scheme is not scalable in its basic version and verification may require many evaluations of H when a prover has previously identified many times. This can lead to a denial of service attacks.

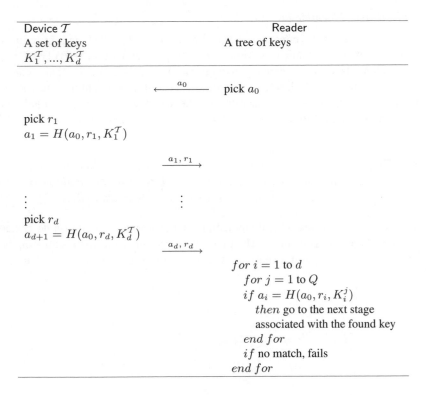

Figure 5.3. *The MSW Identification Protocol*

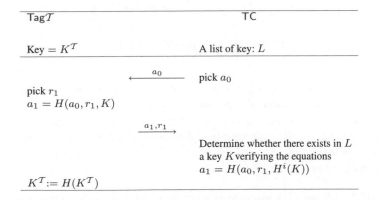

Figure 5.4. *Ohkubo et al. identification protocol*

5.3.1.6. *HB-typed identification protocol*

At the CRYPTO'05 conference Juels and Weis [JUE 05] had introduced a modification of the authentication protocol of Hopper and Blum (HB) [HOP 01] in order to extend its security against active attacks and passive attacks. This scheme is suitable for materials with limited resources because it uses essentially the bit by bit AND and XOR (exclusive OR) operations. The security of these protocols is based on the computational difficulty of the LPN (*Learning Parity with Noise*) problem [BLU 93]:

DEFINITION.– *Let A be a random $k \times n$ binary matrix, x a random n bits vector, $0 < \eta < 1/2$ a noise parameter and v a random n bits vector of weight $wt(v) < \eta n$. Given A, η and $z = (A.x) \oplus v$, find an n bits vector x' such that $wt(A.x' \oplus z) < \eta$.*

The protocol performs several elementary iterations with a iteration described in Figure 5.5 where ν_i is a bit obtained by random source which returns 1 with probability η. The prover and the verifier share the secret key (x, y) and are able to calculate the

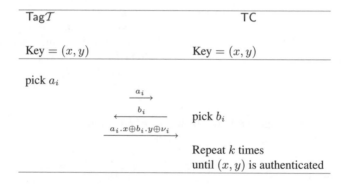

Figure 5.5. *HB$^+$ identification protocol*

value $a_i.x \oplus b_i.y$ from (a_i, b_i) where . represents the inner product modulo 2. The prover sends a noisy version with probability η of this value at each iteration and the verifier must check that the equality with the value that it calculates is true with a number of times higher than a determined threshold. The key should be long enough to ensure its confidentiality. x of 80 bits, y of 512 bits and 1,164 iterations is an example of recommended parameters.

THEOREM.– *If the LPN problem is hard then the scheme is secure against active narrow impersonation attack and narrow-weak private.*

Note that in this context the security against active attacks is only valid for an adversary without access to the VERIFY function (i.e. for example without having a valid reader). Otherwise, a linear attack in the size x was described by Gilbert *et al.* in [GIL 08]. The MITM attack consists in inverting the value of a bit of a_i (resp. b_i), always at the same position along the iterations.

If this operation does not alter the identification result, the position corresponds to a good probability to have a bit at 0 in x (resp. y). This attack allows the impersonation or the recognition of a prover.

To counter this attack, various ideas have been proposed, such as by removing the linearity of the scheme (see HB^{++} [BRI 06b]), by replacing the scalar product by a matrix product and a_i, b_i by binary vectors to counter partially the attacks by the middle (see HB# [GIL 08]) or by adding a final integrity verification step (see Trusted-HB [BRI 08a]). These schemes continue to use elementary operations but the number of iterations required for HB-typed schemes represents a significant drawback.

5.3.2. *Asymmetric cryptography-based protocols*

In general, asymmetric cryptography requires an important computing power, but it allows us to have protocols in which the security depends on established computational assumptions.

The use of asymmetric cryptographic techniques has an additional advantage, it allows us to trace the identity of a tag in a constant time. We now present several identification protocols of this type, which are perfectly scalable.

5.3.2.1. *Computational assumption*

This section briefly recalls the assumptions ensuring the security and the privacy of the schemes presented below. We will focus only on schemes using variations of the discrete logarithm problem. Let $\mathbb{G}$ be a cyclic group, multiplicatively, of order – possibly unknown to the adversary – q, and g a generator element.

The discrete logarithm problem (DL) is as follows: given g and g^a in $\mathbb{G}$, compute a where a is a random element of $[0, ..., q - 1]$.

The discrete logarithm problem with short exponent (DLSE), introduced in [OOR 96], is the same problem restricted to a small exponent a. If a is between 0 and $S - 1$ for S smaller than q, we talk about the DLSE(S) problem.

The computational Diffie-Hellman (CDH) problem can be defined as follows. Given g^a and g^b, compute g^{ab}, where a and b are random elements of $[0, q - 1]$. With small exponents, we talk about the SECDH problem (see [KOS 04]). The difficulty of the CDH problem is that it needs a stronger assumption than DL: if there is an algorithm capable of solving DL, then there exists an algorithm that can solve CDH.

Finally, the Decisional Diffie-Hellman (DDH) problem is defined: given g^a, g^b and g^c, decide if g^{ab} and g^c are equal. We can note that if DDH is difficult, so it is CDH. The case with small exponents SEDDH is also defined in [KOS 04].

It should be noted that the discrete logarithm problem, if it is hard, imposes difficulties on calculating the value of the logarithm. Therefore it is usually possible to ensure that an adversary cannot decipher the secret (this is also true for the CDH). The DDH problem ensures that it is not possible to distinguish a random value and the value g^{ab}. This allows us, in particular for privacy, to prove that it is not possible to distinguish between simulated values and valid transcripts.

5.3.2.2. *Implementation of cryptosystems based on elliptic curves*

An elliptic curve is a plane curve defined by an equation – so-called Weierstrass – of the form:

$$y^2 = x^3 + ax + b.$$

One can note that the set of points (x, y) in which x and y stay on the curve forms a group.

Elliptic curves seem to be the mathematical tool that has the maximum chance to be implemented on RFID tags. In fact, calculations can be performed on relatively small elements because the resolution of the problem of discrete logarithm in the group of points of an elliptic curve is exponentially difficult.

It would be too long and out of scope to explain in greater detail the different possible calculations on elliptic curves. However since 2006, several implementations [WOL 05, KUM 06, LEE 07, FUR 07] have been made to assess the possibility of applying the elliptic curves cryptography in a restricted environment like an RFID tag.

Recent implementations [HEI 08, BOC 08, LEE 08] of the scheme use elliptic curves which require a physical space from 10000 to 15000 GEs. A summary of the characteristics of these implementations is presented in Table 5.2. If we compare to other functions based on symmetric cryptography, this size is still important. However, these results give a hope for the use of elliptic curves on RFID tags in the future.

Propositions	Space (GE)	Power (μW)	Mean Current (μA)
ECC [HEI 08]	11904	10.8	6
ECC [BOC 08]	10392	46	?
ECC [LEE 08]	12168	51.85	?

Table 5.2. *Differents elliptic curve implementations*

These implementations of calculations on elliptic curves can be used for all protocols that are presented in the following sections.

5.3.2.3. *State-of-the-art examples not respecting the privacy*

5.3.2.3.1. Identification protocol of Schnorr

One of the best known zero-knowledge identification schemes is that proposed by Schnorr [SCH 89]. The principle is described by Figure 5.6 where the operations are done either in the group $\mathbb{G}$, or modulo the group order to calculate y.

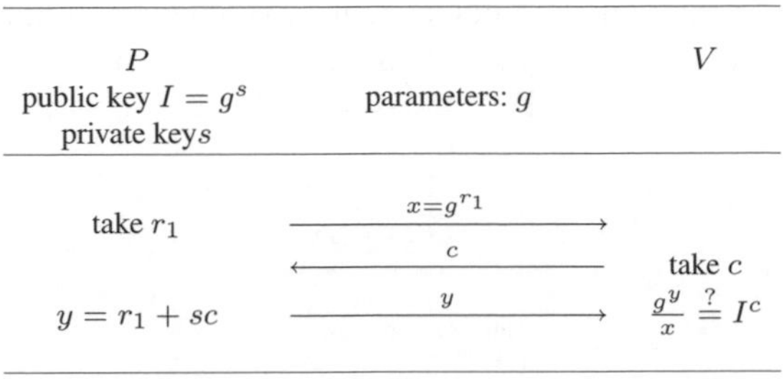

P		V
public key $I = g^s$	parameters: g	
private key s		

take r_1	$\xrightarrow{\quad x=g^{r_1} \quad}$	
	$\xleftarrow{\quad c \quad}$	take c
$y = r_1 + sc$	$\xrightarrow{\quad y \quad}$	$\frac{g^y}{x} \overset{?}{=} I^c$

Figure 5.6. *Identification scheme of Schnorr*

If the CDH problem is hard, then the scheme of Schnorr is secure against passive attacks and it is zero-knowledge for a honest verifier. Note that the scheme is also proved secure against some active attackers [BEL 02] but in a model which is less general than the one defined in section 5.2.2.2).

Many variations exist, including the GPS scheme suggested in [MON 07] as a solution for passports. However, these schemes were not designed with a concern for privacy and although *a priori* adequate by their ZK property, these schemes do not respect the privacy.

When a prover P executes the protocol to prove the knowledge of secret s related to its identity $I = g^s$, then an observer can memorize $g^{r_1}, c, y_1 = r_1 + sc$, which allows the latter to calculate $g^{y_1} g^{-r_1} = g^{sc} = I^c$. Similarly, if another prover P' is identified, the observer obtains $I'^{c'}$, which allows it to verify if I and I' are equal or not. Thus one observer has enough information to distinguish the identity of the prover. This weakness is spread to other identification schemes where an algebraic relationship is used in the proof, such as GPS, Fiat-Shamir [FIA 86], GQ [GUI 88] or other generalizations (GQ2 [QUI 00], Ong-Schnorr [ONG 90], Okamoto [OKA 92], a modification of Fiat-Shamir [MIC 88]).

5.3.2.3.2. GPS identification protocol

The GPS scheme described in [GIR 06] is a very effective ZK identification scheme. It uses the principle of the scheme of Schnorr but by restricting the use of small exponents. It allows the reduction of computing cost of exponentiation and modular reductions in RSA rings or finite fields, and avoids some modular reductions. On elliptic curves, an efficient implementation using pre-computation tokens is possible [GIR 06]. The paper [MCL 07] proposes an implementation of this scheme for RFID.

The protocol (Figure 5.7) is as follows. The prover randomly picks an exponent $r_1 \in [0, A-1]$, computes $x = g^{r_1}$ and sends this value to the verifier. The verifier sends a random challenge $c \in [0, B-1]$ and the prover replies with $y = r_1 + sc$. This calculation is done without modular reduction, contrary to the case of Schnorr. The verifier then verifies the presence of an identity $I \in L$ such that $g^y x^{-1} = I^c$ and $0 \leq y \leq A - 1 + (B-1)(S-1)$. If the two conditions are verified, then the prover is identified.

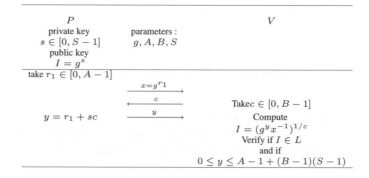

Figure 5.7. *GPS identification protocol*

Its security is based on the DLSE problem. If this problem is hard and if $\frac{BS}{A}$ is negligible, then the GPS scheme is secure against passive attacks and zero-knowledge for a honest verifier.

5.3.2.4. *Adaptations for the privacy*

As we have seen, it is necessary to provide additional protections to these schemes for the privacy purpose. The modifications described below lead to efficient scalable and private schemes. In the following, we assume that $\frac{BS}{A}$ is a negligible quantity.

5.3.2.4.1. Hashed GPS

Instead of computing g^{r_1} in each execution of the protocol, it is possible to work with $H(g^{r_1})$ (see [GIR 06]), where H is a hash function. This idea allows the use of token and the limitation of computing power for a prover, for example, by pre-calculating these values and storing the pairs $(r_1, H(g^{r_1}))$ as usable one time tokens. In that case, the prover has to calculate $r_1 + sc$ only. It is easy to implement even without modular reduction. Practically, r_1 can be generated by a recursive pseudo-random function using a particular seed, it is sufficient to store the initial seed and update it every time. Thus the size of a token is about the size of a hash. For example for a hash of 50 bits in length, we can store about 640 tokens in a memory of 4 KB. Moreover, this modification will improve the privacy of the scheme.

THEOREM.– *If the problem DLSE(S) is hard and if H is pre-image resistant then Hashed GPS is secure against active attacks, ZK for a honest verifier and weak-private with hidden identity.*

This modification makes it possible to consider an implementation with privacy for low cost RFID tags, but its disadvantage is the non extensibility to the use of multiple tags.

5.3.2.4.2. Randomized GPS

The Randomized GPS scheme is described in Figure 5.8. In this scheme, the privacy is ensured by introducing a pair of asymmetric keys for the verifier. We assume that the prover knows the verifier's public key.

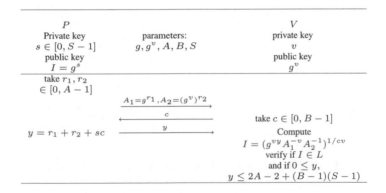

Figure 5.8. *Randomized GPS identification protocol*

We obtain the following result:

THEOREM.– *Assume SEDDH is hard, the Randomized GPS scheme is secure against passive eavesdroppers, to zero knowledge disclosure and PI narrow-strong private.*

The main difference compared to the GPS scheme comes from the calculation A_2^v ensuring that only the verifier can make the final verification, and thus distinguish a valid transcript from a simulated one.

This scheme is fully scalable, since the verifier performs a constant number of operations to identify the prover, i.e. without slowing down the identification process regardless of the number of provers and identities in the system.

Note that [BAT 06] describes an implementation on an area of about 13,000 GEs based on elliptic curves of the scheme of Schnorr in a low-cost equipment. Its architecture can implement the Randomized GPS scheme, but with an additional property of privacy.

5.3.2.4.3. Randomized Hashed GPS

This scheme provides us security against active attacks and maintain the privacy even in cases of corruption. The difference compared to the previous case comes from the application of a hash function to the first message: it is somehow a combination of the two previous ideas. Details are provided in Figure 5.9.

P
private key
$s \in [0, S-1]$
public key
$I = g^s$

parameters:
g, g^v, A, B, S

V
private key
v
public key
g^v

take r_1, r_2
in $[0, A-1]$

$$z = H\left(g^{r_1}, (g^v)^{r_2}\right) \longrightarrow$$

$$\longleftarrow c$$

take $c \in [0, B-1]$

$$y = r_1 + r_2 + sc \qquad A_1 = g^{r_1}, A_2 = (g^v)^{r_2}, y \longrightarrow$$

Compute
$$I = (g^{vy} A_1^{-v} A_2^{-1})^{1/cv}$$
Verify if $I \in L$
and $z = H(A_1, A_2)$
and if $0 \le y$,
$y \le 2A - 2 + (B-1)(S-1)$

Figure 5.9. *Randomized Hashed GPS identification protocol*

Here several interesting properties appear: all calculations can be done off-line, and do not reveal information about the secrecy of the prover. This scheme is also fully scalable.

THEOREM.– *Assume the problem SEDDH is hard and H is pre-image and collision resistant. Then Randomized Hashed GPS is PI forward private, secure against active attacks, PI narrow-strong private and ZK.*

As the scheme is now public-identity forward private, it could be used in common applications where identities are public, for example for identity documents like passports. Moreover in this case, the list of authorized identities can be reduced to the identity of the passport in verification which is directly obtained by scanning the document itself, eliminating the problem of storage of the list on the verifier's side.

5.3.3. *Protocols based on physical properties*

In the previous sections, the security of cryptographic tools is based on algorithmic assumptions, i.e. there is no efficient algorithm to solve some algorithmic problems, such as finding a collision or a discrete logarithm. In this section, we present cryptographic tools where the security is based on physical properties, PUF (*Physical Unclonable Function*) functions and POK (*Physically Obfuscated Key*) keys.

Other interesting solutions use physical properties to improve the security of cryptographic protocols for RFID. [BIR 07] gives an overview of some protocols combining algorithmic processes and physical principles. The advantage of this type of solution is that it requires much fewer resources in tag than a purely cryptographic solution. Solutions with restricted uses were first proposed (such as tag deactivation, utilisation of a removable antenna, a radio "shield", distance limitations, etc.). For example, in [JUE 03], the principle of *blocker tag* was proposed to prevent a reader from recognizing some RFID tags. The blocker tag is an additional component simulating multiple tags responding simultaneously to readers in parallel with other surrounding tags, so as to confuse the identification of the latter ones. It can be configured to protect only a subset of tags and thus improves the privacy by blocking the reading of selected tags.

More recently, techniques exploiting the noise of the communication channel are also highlighted. The fact of exploiting the channel noise is not a new idea because this is one of the foundations of quantum information theory. However, its application to the context of RFID is a new idea. Thus, [CHA 06] proposes a protocol for key establishment well suited to tags with limited resources by developing an algorithm based on the principle of distillation-reconciliation-amplification. The noise of the channel creates an interference against a passive attacker. In the case of passive tags behaving like memories with very low computing capacity, the insertion of an artificial interference which the adversary cannot subtract (principle of *Noisy Tag* and *Noisy Reader*) is also proposed to protect the establishment or exchange of a secret between a reader and a tag in the presence of possible channel eavesdropping [CAS 06, BRI 06a, SAV 07].

In this last part, we will introduce first the definition of the PUF (Physical Function Unclonable) functions and then that of the POK (Physically Obfuscated Key) keys. Finally, we present two schemes using PUF or POK and modifying schemes of section 5.3.1, the HB protocol and the MSW protocol. In fact, the use of PUF aims at finding a substitute for asymmetric principles in order to improve security and privacy at a lower cost.

5.3.3.1. *Definition of PUFs and POKs*

RFID chips, particularly low cost ones, have risks of leaking the values they contain, or risks of cloning. To be protected against them, the use of PUF (Physical Unclonable Function) to ensure the physical protection of an electronic component is studied. The aim is to associate and physically protect the secrets contained in an electronic component, so that a modification in the integrity of the component, for example, an attempt to access the memory results in an irremediable alteration of these secrets.

In his thesis [RAV 01], Pappu introduced a notion of one-way physical function. The possibility of considering this type of function is due to the random nature of some physical properties of a component during its manufacture. The example used in [RAV 01] is the random incorporation of translucent bubbles in a resin. The distribution of these bubbles can then generate a data that is intrinsically linked to the object containing the resin.

In 2003, Gassend [GAS 03] proposed an extension of the concept of PUF to ensure the non-clonability of a component on the same principle. A PUF is a function associating a response to a stimulation such as:

1) the function is easy to evaluate;

2) it is difficult to characterize by physical observation or from known couples (stimulation, response);

3) it is difficult to regenerate exactly.

Despite its physical character, a PUF must keep the same behavior (i.e. same stimulation-response values) at an acceptable change in operating conditions (heat, humidity, etc.).

The difficulty of characterizing a PUF can be interpreted as an impossibility of characterization using a polynomial amount of resources (money, time, etc.). A PUF is then often modelled as a pseudo-random function[1]. To simplify the analysis, we assume in the following part of the chapter that a PUF is a truly random function. Compared to the definition of Gassend [GAS 03], the notion of I-PUF (Integrated PUF) defined in [TUY 06] requires the following additional properties:

1) the PUF is irretrievably linked to the chip (or component) using it. In particular, any attempt to separate them destroys the PUF chip, it is impossible to obtain information on communications between the chip and the PUF;

2) an attacker cannot get the output value of the PUF.

These properties are intended to protect the component against attempts of physical attacks since no information on an I-PUF can be obtained. Thereafter, if necessary,

---

1 It is possible that the number of possible stimulations is bounded.

PUFs used in the described protocols will be chosen as I-PUF to ensure the resistance of RFID tags against physical attacks.

A so-called silicon-type PUF as described in [PSM 02] is a good example of I-PUF. For an integrated circuit in silicon, it is obtained by comparisons between two random rows of the circuit, and propagation delays of a signal. This feature depends on the final circuit because from an electronic component to another, the inherent irregularities in the manufacture of semiconductors implies the apparition of variable delays especially during transitions of a logic gate. This implementation requires very few resources and a practical application of RFID is described in [SUH 07]. Companies have recently commercialized RFID tags using similar PUFs. Concretely, given a binary input value, the circuit in which the delay will be measured, is unique. Figure 5.10 summarizes this idea.

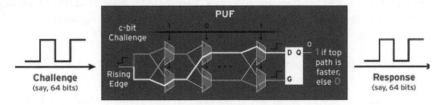

Figure 5.10. *Example of PUF*

The concept of POK (Physically Obfuscated Key) is also given by Gassend in [GAS 03]. This key is contained in a component. An external adversary having a polynomial amount of resources will not be able to locate the key. It can be interpreted as a PUF answering only to a single stimulation that cannot have a fixed answer. An example of implementation (see Figure 5.11) is proposed in [PSM 03] from a PUF, a determined stimulation and a hard-wired fuse of constant value *Fuse*. To create a POK of a binary vector K, one need only set a stimulation in the input of the PUF and choose the value *Fuse*, so that bit by bit is added in the response of the PUF, the result is equal to K.

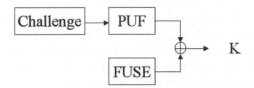

Figure 5.11. *Example of POK*

Thanks to the properties of the PUF, which behaves like a perfect random function, the knowledge of its input and the value *Fuse* provides no information on the key.

This concept has an advantage for an RFID tag: a POK can be considered as a value stored by the tag, but which is only accessible after the stimulation of the associated PUF.

5.3.3.2. *How can we use POKs?*

A key K involved in a cryptographic calculation must be stored in a memory area at a given moment in order to be used. Such a memory area is not necessarily secure, especially in the case of an RFID tag. Therefore, despite the use of a POK, other precautions are necessary to prevent an attacker from directly reading the memory in a tag, so that the key K is not revealed to the attacker. This problem has been studied in [GAS 03] with the proposal of a general protection: splitting the calculations and the key in two parts and use them one after another. The idea is close to the secret masking used on embedded systems to counter against side channel attacks such SPA (Simple Power Analysis). These methods are generally constructed via homomorphic relations, for example for an exponentiation: if $K = K' \times K''$, $z = g^K$ is computed by $y = g^{K'}$ then $z = y^{K''}$.

In the case of the protocol described in section 5.3.3.4, the computations use bit to bit additions ($\oplus$) and the "splitting" of K is done as follows:

1) a sub-key K' is a value randomly chosen and then implemented by a POK;

2) $K'' = K \oplus K'$ is computed and the value K'' is then implemented by a second POK.

In this way, regardless of the moment when the memory is attacked, no information is obtained about the key, since (1) at most one of the two halves of the key is accessible in memory, and (2) after attack, the properties of underlying PUFs make the tag unusable. On the other hand, the splitting of the key K is randomly performed, even if the same key is contained in several RFID tags, the disclosure of memory does not allow an attacker to obtain information about K since each splitting is different.

5.3.3.3. *PUF-HB*

The scheme defined in [HAM 08b] uses a PUF to mask the value of the calculation made by the prover. The basic principle is: instead of returning a value of type $b_i.y \oplus \nu_i$ (for the original version[2] of HB), the prover returns a value of type $(b_i, a_i.y \oplus \nu_i \oplus PUF_K(b_i))$ where a_i is randomly taken by the prover and b_i is a stimulation for which the verifier has learned the corresponding response before the release of the prover. Thus, only the verifier is able to verify the received value.

During its construction [HAM 08b], a PUF is implemented by the comparison of the propagation time of a signal through different sub-circuits. The algebraic model

---

2 The additional term $a_i.x$ was inserted in HB$^+$ to counter malicious verifiers.

used to represent the phenomena is as follows. Given a stimulation $a = (a_1, ..., a_l)$, p_i is defined by the value $\oplus_i^l a_j$. Given K a key of l bits $k_1, ..., k_l$, the response of RUF is:

$$PUF_K(a) = sign\left(\sum_{i=1}^{l-1}(-1)^{p_i}k_i \oplus k_l\right) \text{ if } a_l.k_l = 1 \qquad [5.1]$$

$$PUF_K(a) = sign\left(\sum_{i=1}^{l-1}(-1)^{\bar{p}_i}k_i \oplus k_l\right) \text{ if } a_l.k_l = 0 \qquad [5.2]$$

where $sign(x) = \frac{x}{|x|}$. The results given in [HAM 08b] is:

THEOREM.– *Assume the LPN problem is hard, then the HB protocol is narrow weak private and secure against a narrow active attacker.*

If the function PUF is perfect, then a strong adversary is equivalent to a weak adversary. Note that the use of PUF is only designed to protect against compromises of tags and the possibility to perform an attack of type MITM for HB$^+$ remains.

A similar modification is proposed in [HAM 08a] on the HB$^+$ protocol to simplify the security proof. A more accurate implementation is described for which the required area is estimated at 960 gates (GEs), which is an interesting candidate for RFID.

5.3.3.4. *POK-MSW*

[BRI 08b] presents a modification of the MSW scheme using the principle of POK to improve the privacy. The original scheme is adapted, so that they can use the relation $K = K' \oplus K''$, corresponding to the implementation of K by two POK, in two successive calculations. The scheme is described in Figure 5.12.

Hence, in the reception of a challenge a_0, instead of calculating $H(a_0, r_0, K)$ and returning $(H(a_0, r_0, K), r_0)$, the prover calculates and returns $(H(a_0, r_0), r_0 \oplus K)$. This can introduce a homomorphic operation and split the calculation of the second term in two steps via $r_0 \oplus K = (r_0 \oplus K') \oplus K''$. The use of POK in this way provides resistance against physical threats: the CORRUPT oracle does not help to obtain information about the identity of the prover.

THEOREM.– *Assume H is pre-image and collision resistant, then the scheme is secure against active impersonation attack and weak private. If in addition, the used POK is perfect, then a strong adversary is a weak adversary, i.e. the scheme is strong private.*

Tag $\mathcal{T}$		TC
take $r_1^{\mathcal{T}}$	$\xleftarrow{\quad a_0 \quad}$	take a_0

$a_1 = H(a_0, r_1^{\mathcal{T}})$

The first POK is activated to have $K_1'^{\mathcal{T}}$
$A' = r_1^{\mathcal{T}} \oplus K_1'^{\mathcal{T}}$

The second POK is activated to have $K_1''^{\mathcal{T}}$

$\begin{aligned} A'' &= A' \oplus K_1''^{\mathcal{T}} \\ &= r_1^{\mathcal{T}} \oplus K_1^{\mathcal{T}} \end{aligned}$ $\xrightarrow{\quad a_1, \, r_1^{\mathcal{T}} \oplus K_1^{\mathcal{T}} \quad}$

$\vdots$ $\vdots$

take $r_d^{\mathcal{T}}$
$a_d = H(a_{d-1}, r_d^{\mathcal{T}})$

$\qquad \qquad for\ i = 1\ to\ d$
$\qquad \qquad \quad for\ j = 1\ to\ Q$
$\qquad \qquad \quad r_i^{j'} = \hat{K}_i^j \oplus r_i^{\mathcal{T}} \oplus K_i^{\mathcal{T}}$
$\qquad \qquad \quad if\ a_i = H(a_{i-1}, r_i^{j'})$

$\vdots$ $\xrightarrow{\quad a_d, \, r_d^{\mathcal{T}} \oplus K_d^{\mathcal{T}} \quad}$ $\qquad \qquad \quad \quad then$ go to next floor
$\qquad \qquad \quad \quad$ associated to the found key
$\qquad \qquad \quad end\ pour$
$\qquad \qquad \quad if$ no key is suitable fail
$\qquad \qquad end\ for$

Figure 5.12. *POK-MSW protocol*

In addition, the extensibility of the protocol allows it to target an efficient implementation for an RFID system.

5.3.4. *Summary*

In this chapter, several authentication protocols for RFID have been presented. All of them provide solutions to the problem of privacy but at different levels, as formally defined in section 5.2.3.2. In addition to this property, it is interesting to compare these protocols in terms of their ability to be used for a large number of RFID, i.e. the property of scaling. In Table 5.3, we provide a list of the different protocols with their cryptographic primitives, the privacy level and their ability or not to be extended.

Protocols	Cryptographic primitives	Privacy	Scalability
WRSE	Hash function	Weak private	No
MSW	Hash function	Weak private	Yes
HB$^+$		Weak private	No
Hashed GPS	Hash function, coupons	hidden identity Weak private	No
Randomized GPS	Elliptic curve	*Narrow* Strong private	Yes
Randomized Hashed GPS	Hash function, elliptic curve	Public identity *Forward* private, *Narrow* strong private	Yes
POK-MSW	POK, hash function	Strong private	Yes
PUF-HB	PUF	*Narrow* strong private	No

Table 5.3. *Overview of protocols*

5.4. Conclusion. Physical attacks on RFID devices

5.4.1. *Side-channel attacks*

Since more than a decade, Side-Channel Analysis (SCA) has drawn the attention of researchers and industries in the world of smart cards. These attacks exploit the dependence between the power consumption, the electromagnetic emanation, the execution time of the card and the executed operations and data. This kind of attack becomes more and more sophisticated, in term of equipment and post-processing with advanced signal processing techniques. Typically, we distinguish simple analysis in which only simple observations are needed (for example, direct measures of power consumption SPA (*Simple Power Analysis*) or electromagnetic SEMA (*Simple ElectroMagnetic Analysis*) and differential analysis which needs a more sophisticated statistical processing on acquired signals (for example DPA (*Differential Power Analysis*) or DEMA (pour *Differential ElectroMagnetic Analysis*).

To respond to these threats, different types of counter-measures have been proposed and implemented to protect the component. Obviously, the countermeasures are not free and they have an impact on the complexity and the performance.

For several years, RFID tags have become a new target for physical attacks. Because of the particular characteristics of the RFID system, the physical measurements are generally more complex than those of a smart card in the contact mode. First, RFID tags, in most cases, are not fed directly by a stable power supply but by the electromagnetic field of an RFID reader. Second, the electromagnetic emission of an RFID tag is superimposed on the RF field of the reader. Therefore specific equipment and processing are needed to extract useful information issued by the tag.

There are two approaches to measure the power consumption of an RFID tag. The first approach uses a resistor placed in one of the power supply lines of the module performing cryptographic operations. The voltage drop across this resistance is proportional to the circulating current and therefore is proportional to the power consumption of the tag. This approach is direct and simple, but it is not easy to achieve during a real attack. In all cases, the results obtained by this method can be used as a reference to evaluate other attack methods [HUT 07].

The second approach tries to estimate indirectly the power consumption of the tag. In [ORE 07], Oren and Shamir presented an attack called *parasitic backscatter attack* applied to UHF RFID tags operating at 900 MHz. The advantage of this attack is the possibility to estimate the power consumption of the RFID tag without touching the tag or the reader. In fact, it is based on the relationship between the power consumption of the tag and the power of the field reflected by the tag to communicate with the reader (backscatter). By measuring the reflected field we can calculate the power consumption of the attacked tag. This measurement is performed using a directive antenna to remove the field emitted by the reader. The paper shows that this attack can extract the secret *Kill Password* of a tag and then enable its definitive deactivation. The considered tags are passive tags of the Class 1, Generation 1 as specified in the standard EPC [INC 08].

Another work [PLO 08a] focuses on electromagnetic measurements and shows that an electromagnetic differential attack is possible on RFID tags located at a distance of one meter from the receiver. An attack, using backscatter signals of the tag, is indeed presented in [PLO 08a] in which off-the-shelf UHF RFID tags are tested. Backscatter signals from the tag are measured by an antenna during the execution of a write-data command. The results show that the written data (2 bytes) is found with 1,000 measurements if the antenna is 20 cm from the tag and with 10,000 measurements if the antenna is at a distance of one meter.

The electromagnetic analysis has obtained successes with smart cards in the contact mode. In the RFID context, measurements of electromagnetic signals are more complex. The presence of the RF field of the reader and the low power of the electromagnetic field of the tag are major difficulties. Some solutions have been proposed by Hutter *et al.* [HUT 07], for example, the use of a receiver to eliminate the carrier of the field emitted by the reader or the use of a Helmholtz arrangement, which is a specified test set-up for compliance testing, as described in the ISO / IEC 10373-6

[ISO 01]. In this work, two RFID 13.56 MHz prototypes implementing the AES block cipher in software and hardware are analyzed. The result shows that the DEMA attack is successful with about 1,000 measurements.

Post-processing techniques to improve the attack effectiveness are also applied to RFID tags. An example is the transformation of time signals in the frequency domain to reduce the effects caused by the synchronization. This technique has been tested with success on contact-mode smart card and can also be applied to RFID tags as presented in [PLO 08b].

5.4.2. *Fault injection attacks*

The fault injection attacks can allow an attacker to obtain secret information by disrupting the execution of a cryptographic application. Different types of fault can be considered: clock, temperature, voltage or laser. While fault attacks on smart cards have achieved real progress, the paper [HUT 08] can be considered as the first publication of this kind of attack on RFID. Faults in power consumption and clock can be achieved temporarily by interconnecting of the pins of the tag's antenna. In this way, the antenna was suspended for a time period and changes in consumption and clock were made. Electromagnetic injections were performed with the help of a high voltage generator which is capable of producing electromagnetic pulses. The laser attacks are possible at both global level (entire surface of the chip) and local one (specific positions). Experiments show that RFID tags are vulnerable to faults during the writing of data stored in the internal memory.

5.4.3. *KeeLoq*

The previous analysis shows the vulnerability of RFID tags to physical attacks. Counter-measures are thus necessary. However, RFID tags, with particular characteristics of size, performance and consumption, making it very difficult to implement the counter-measures proposed to smart cards.

To conclude this chapter, we give an example showing that the ignorance of this type of protection could cause of heavy consequences.

KeeLoq is an RFID system that uses a symmetric encryption. Although KeeLoq was designed in the 1980s, the first attack was published in 2007 by Bogdanov [BOG 07b, BOG 07a]. The attack – such as those presented in the introduction – consists of two steps. Initially, the proprietary algorithm was found through reverse engineering techniques. In a second step, the mathematical weakness of the underlying cryptographic algorithm has been exploited to undermine the system.

What is remarkable here is the apparition of the third step with the publication at the Crypto 2008 conference [EIS 08] about a differential power consumption analysis on KeeLoq tags. This physical attack was revealed to be more devastating thus making it possible to completely break the KeeLoq system.

5.5. Bibliography

[AVO 07] AVOINE G., KALACH K., QUISQUATER J.-J., "Belgian Biometric Passport does not get a pass...", http://www.dice.ucl.ac.be/crypto/passport/index.html, 2007.

[BAT 06] BATINA L., GUAJARDO J., KERINS T., MENTENS N., TUYLS P., VERBAUWHEDE I., "An elliptic curve processor suitable for RFID tags", *WISSEC*, Belgium, November 8-9 2006.

[BEL 02] BELLARE M., PALACIO A., "GQ and Schnorr identification schemes: proofs of security against impersonation under active and concurrent attacks.", YUNG M. (ed.), *Advances in Cryptology - CRYPTO 2002, 22nd Annual International Cryptology Conference, Santa Barbara, California*, USA, August, Springer, p. 162-177, 2002.

[BIR 07] BIRD N., CONRADO C., GUAJARDO J., MAUBACH S., SCHRIJEN G.J., SKORIC B., TOMBEUR A.M.H., THUERINGER P., TUYLS P., "ALGSICS - combining physics and cryptography to enhance security and privacy in RFID systems", STAJANO F., MEADOWS C., CAPKUN S., MOORE T. (eds), *ESAS*, vol. 4572 of *Lecture Notes in Computer Science*, Springer, p. 187-202, 2007.

[BLU 93] BLUM A., FURST M.L., KEARNS M.J., LIPTON R.J., "Cryptographic primitives based on hard learning problems", STINSON D. R. (ed.), *CRYPTO*, vol. 773 of *Lecture Notes in Computer Science*, Springer, p. 278-291, 1993.

[BOC 08] BOCK H., BRAUN M., DICHTL M., HESS E., HEYSZL J., KARGL W., KOROSCHETZ H., MEYER B., SEUSCHEK H., "A milestone towards RFID products offering asymmetric authentication based on elliptic curve cryptography", *Invited Talk in Workshop on RFID Security, RFIDSec2008*, Budapest, 2008.

[BOG 07a] BOGDANOV A., "Attacks on the KeeLoq Block Cipher and Authentication Systems", *Conference on RFID Security*, Malaga, Spain, July, 2007.

[BOG 07b] BOGDANOV A., 'Cryptanalysis of the KeeLoq block cipher', Cryptology ePrint Archive, Report 2007/055, 2007, http://eprint.iacr.org/.

[BOG 07c] BOGDANOV A., KNUDSEN L.R., LEANDER G., PAAR C., POSCHMANN A., ROBSHAW M.J.B., SEURIN Y., VIKKELSOE C., "PRESENT: An Ultra-Lightweight Block Cipher", in: PAILLIER P., VERBAUWHEDE I. (eds), Cryptographic Hardware and Embedded Systems - CHES 2007, 9th International Workshop, Vienna, Austria, p. 450-466, 2007.

[BRI 06a] BRINGER J., CHABANNE H., "On the wiretap channel induced by noisy tags", BUTTYÁN L., GLIGOR V.D., WESTHOFF D. (eds), *ESAS*, vol. 4357 of *Lecture Notes in Computer Science*, Springer, p. 113-120, 2006.

[BRI 06b] BRINGER J., CHABANNE H., DOTTAX E., "HB^{++}: a Lightweight authentication protocol secure against some attacks", *SecPerU*, IEEE Computer Society, p. 28-33, 2006.

[BRI 08a] BRINGER J., CHABANNE H., "Trusted-HB: a low-cost version of HB$^+$ secure against man-in-the-middle attacks", *IEEE Transactions on Information Theory*, vol. 54, no. 9, p. 4339-4342, 2008.

[BRI 08b] BRINGER J., CHABANNE H., ICART T., "Improved privacy of the tree-based hash protocols using physically unclonable function", OSTROVSKY R., PRISCO R. D., VISCONTI I. (eds), *SCN*, vol. 5229 of *Lecture Notes in Computer Science*, Springer, p. 77-91, 2008.

[CAS 06] CASTELLUCCIA C., AVOINE G., "Noisy tags: a pretty good key exchange protocol for RFID tags", DOMINGO-FERRER J., POSEGGA J., SCHRECKLING D. (eds), *CARDIS*, vol. 3928 of *Lecture Notes in Computer Science*, Springer, p. 289-299, 2006.

[CHA 06] CHABANNE H., FUMAROLI G., "Noisy cryptographic protocols for low-cost RFID tags", *IEEE Transactions on Information Theory*, vol. 52, no. 8, p. 3562-3566, 2006.

[EIS 08] EISENBARTH T., KASPER T., MORADI A., PAAR C., SALMASIZADEH M., SHALMANI M.T.M., "On the power of power analysis in the real world: A complete break of the KeeLoqCode hopping scheme", WAGNER D. (ed.), *CRYPTO*, vol. 5157 of *Lecture Notes in Computer Science*, Springer, p. 203-220, 2008.

[FEL 04] FELDHOFER M., DOMINIKUS S., WOLKERSTORFER J., "Strong authentication for RFID systems using the AES algorithm", JOYE M., QUISQUATER J.-J., (eds), *CHES*, vol. 3156 of *Lecture Notes in Computer Science*, Springer, p. 357-370, 2004.

[FEL 07a] FELDHOFER M., "Comparison of Low-Power Implementations of Trivium and Grain", *The ECRYPT Stream Cipher Project*, 2007, www.ecrypt.eu.org/stream/papersdir/2007/027.pdf.

[FEL 07b] FELDHOFER M., WOLKERSTORFER J., "Strong crypto for RFID tags - a comparison of low-power hardware implementations", *ISCAS*, IEEE, p. 1839-1842, 2007.

[FIA 86] FIAT A., SHAMIR A., "How to Prove Yourself: Practical Solutions to Identification and Signature Problems.", ODLYZKO A.M. (ed.), *Advances in Cryptology - CRYPTO '86*, Santa Barbara, California, USA, Springer, p. 186-194, 1986.

[FUR 07] FURBASS F., WOLKERSTORFER J., "ECC processor with low die size for RFID applications", *IEEE International Symposium on Circuits and Systems ISCAS 2007*, New Orleans, p. 1835-1838, 2007.

[GAR 08] GARCIA F.D., DE KONING GANS G., MUIJRERS R., VAN ROSSUM P., VERDULT R., SCHREUR R. W., JACOBS B., "Dismantling MIFARE Classic", JAJODIA S., LÓPEZ J., (eds), *ESORICS*, vol. 5283 of *Lecture Notes in Computer Science*, Springer, p. 97-114, 2008.

[GAS 03] GASSEND B., Physical Random Functions, Master's thesis, Computation Structures Group, Computer Science and Artificial Intelligence Laboratory, Massachusetts Institute of Technology, 2003.

[GIL 08] GILBERT H., ROBSHAW M.J.B., SEURIN Y., "HB$^{\#}$: Increasing the Security and Efficiency of HB$^+$", SMART N.P. (ed.), *EUROCRYPT*, vol. 4965 of *Lecture Notes in Computer Science*, Springer, p. 361-378, 2008.

[GIR 06] GIRAULT M., POUPARD G., STERN J., "On the Fly Authentication and Signature Schemes Based on Groups of Unknown Order", *J. Cryptology*, vol. 19, num. 4, p. 463-487, 2006.

[GRI 08] GRIMAUD G., STANDAERT F.-X. (eds), *Smart Card Research and Advanced Applications, 8th IFIP WG 8.8/11.2 International Conference, CARDIS 2008, London, UK*, vol. 5189 of *Lecture Notes in Computer Science*, Springer, 2008.

[GUI 88] GUILLOU L.C., QUISQUATER J.-J., "A "paradoxical" identity-based signature scheme resulting from zero-knowledge.", GOLDWASSER S. (ed.), *Advances in Cryptology - CRYPTO '88, 8th Annual International Cryptology Conference*, Santa Barbara, California, USA, August 21-25, Springer, p. 216-231, 1988.

[HAM 08a] HAMMOURI G., ÖZTÜRK E., BIRAND B., SUNAR B., "Unclonable lightweight authentication scheme", CHEN L., RYAN M.D., WANG G. (eds), *ICICS*, vol. 5308 of *Lecture Notes in Computer Science*, Springer, p. 33-48, 2008.

[HAM 08b] HAMMOURI G., SUNAR B., "PUF-HB: a tamper-resilient HB based authentication protocol", BELLOVIN S.M., GENNARO R., KEROMYTIS A.D., YUNG M. (eds), *ACNS*, vol. 5037 of *Lecture Notes in Computer Science*, p. 346-365, 2008.

[HAN 05] HANCKE G., A practical relay attack on ISO 14443 proximity cards, http://www.cl.cam.ac.uk/gh275/relay.pdf, 2005.

[HEI 08] HEIN D., WOLKERSTORFER J., N.FELBER, "ECCon: ECC is ready for RFID - a proof in silicon", *Workshop on RFID Security, RFIDSec2008*, Budapest, 2008.

[HON 06] HONG D., SUNG J., HONG S., LIM J., LEE S., KOO B., LEE C., CHANG D., LEE J., JEONG K., KIM H., KIM J., CHEE S., "HIGHT: a new block cipher suitable for low-resource device", GOUBIN L., MATSUI M. (eds), *CHES*, vol. 4249 of *Lecture Notes in Computer Science*, Springer, p. 46-59, 2006.

[HOP 01] HOPPER N.J., BLUM M., "Secure Human Identification Protocols", BOYD C. (ed.), *ASIACRYPT*, vol. 2248 of *Lecture Notes in Computer Science*, Springer, p. 52-66, 2001.

[HUT 07] HUTTER M., MANGARD S., FELDHOFER M., "Power and EM Attacks on Passive 13.56 MHz RFID Devices", in: PAILLIER P., VERBAUWHEDE I. (eds), Cryptographic Hardware and Embedded Systems - CHES 2007, 9th International Workshop, Vienna, Austria, p. 320-333, 2007.

[HUT 08] HUTTER M., SCHMIDT J.-M., PLOS T., "RFID and its vulnerability to faults", OSWALD E., ROHATGI P. (eds), *CHES*, vol. 5154 of *Lecture Notes in Computer Science*, Springer, p. 363-379, 2008.

[INC 08] INC. E., UHF Class 1 Gen 2 Standard v. 1.2.0, 2008, http://www.epcglobalinc.org/standards/uhfc1g2/uhfc1g2_1_2_0-standard-20080511.pdf.

[ISO 01] ISO/IEC 10373-6: Identification cards - Test methods - Part 6: Proximity cards, 2001.

[JUE 03] JUELS A., RIVEST R.L., SZYDLO M., "The blocker tag: selective blocking of RFID tags for consumer privacy", JAJODIA S., ATLURI V., JAEGER T. (eds), *ACM Conference on Computer and Communications Security*, ACM, p. 103-111, 2003.

[JUE 05] JUELS A., WEIS S.A., "Authenticating Pervasive Devices with Human Protocols", SHOUP V. (ed.), *CRYPTO*, vol. 3621 of *Lecture Notes in Computer Science*, Springer, p. 293-308, 2005.

[JUE 07] JUELS A., WEIS S.A., "Defining Strong Privacy for RFID", *PERCOMW '07: Proceedings of the Fifth IEEE International Conference on Pervasive Computing and Communications Workshops*, IEEE Computer Society USA, p. 342–347, 2007, http://saweis.net/pdfs/JuelsWeis-RFID-Privacy.pdf.

[KER] KERCKHOFFS A., "La cryptographie militaire", *Journal des sciences militaires*, vol. IX, p. 5-83, January 1883, p. 161-191, February 1883.

[KON 08] DE KONING GANS G., HOEPMAN J.-H., GARCIA F.D., "A practical attack on the MIFARE classic", GRIMAUD G., STANDAERT F.-X. (eds) Smart Card Research and Advanced Applications, 8th IFIP WG 8.8/11.2 International Conference, CARDIS 2008, London, p. 267-282, 2008.

[KOS 04] KOSHIBA T., KUROSAWA K., "Short exponent diffie-hellman problems.", BAO F., DENG R.H., ZHOU J. (eds), *Public Key Cryptography - PKC 2004, 7th International Workshop on Theory and Practice in Public Key Cryptography, Singapore, March, 2004*, Springer, p. 173-186, 2004.

[KUM 06] KUMAR S., PAAR C., "Are standards compliant elliptic curve cryptosystems feasible on RFID?", *Workshop on RFID Security, RFIDSec 2006*, Graz, 2006.

[LE 07] LE T.V., BURMESTER M., DE MEDEIROS B., "Universally composable and forward-secure RFID authentication and authenticated key exchange.", BAO F., MILLER S. (eds), *Proceedings of the 2007 ACM Symposium on Information, Computer and Communications Security, ASIACCS 2007, Singapore, March, 2007*, ACM, p. 242-252, 2007.

[LEA 07] LEANDER G., PAAR C., POSCHMANN A., SCHRAMM K., "New lightweight DES variants", BIRYUKOV A. (ed.), *FSE*, vol. 4593 of *Lecture Notes in Computer Science*, Springer, p. 196-210, 2007.

[LEE 07] LEE Y., VERBAUWHEDE I., "A compact architecture for Montgomery elliptic curve scalar multiplication processor", *International Workshop on Information Security Applications (WISA)*, South Korea, 2007.

[LEE 08] LEE Y., SAKIYAMA K., BATINA L., VERBAUWHEDE I., "A compact ECC processor for pervasive computing", *Workshop on Secure Component and System Identification (SECSI)*, Berlin, 2008.

[LIM 05] LIM C.H., KORKISHKO T., "mCrypton - a lightweight block cipher for security of low-cost RFID tags and sensors", SONG J., KWON T., YUNG M. (eds), *WISA*, vol. 3786 of *Lecture Notes in Computer Science*, Springer, p. 243-258, 2005.

[MAR 05] MARTIN FELDHOFER J.W., RIJMEN V., "AES implementation on a grain of sand", *IEE Information Security*, p. 13-20, 2005.

[MCL 07] MCLOONE M., ROBSHAW M.J.B., "Public key cryptography and RFID tags", ABE M. (ed.), *CT-RSA*, vol. 4377 of *Lecture Notes in Computer Science*, Springer, p. 372-384, 2007.

[MEN 96] MENEZES A., VAN OORSCHOT P.C., VANSTONE S.A., *Handbook of Applied Cryptography*, CRC Press, 1996.

[MIC 88] MICALI S., SHAMIR A., "An improvement of the Fiat-Shamir identification and signature scheme.", GOLDWASSER S. (ed.), *Advances in Cryptology - CRYPTO '88, 8th Annual International Cryptology Conference, Santa Barbara*, USA, Springer, p. 244-247, August 1988.

[MOL 04] MOLNAR D., WAGNER D., "Privacy and security in library RFID: issues, practices, and architectures", ATLURI V., PFITZMANN B., McDANIEL P.D. (eds), *ACM Conference on Computer and Communications Security*, ACM, p. 210-219, 2004.

[MOL 05] MOLNAR D., SOPPERA A., WAGNER D., "A scalable, delegatable pseudonym protocol enabling ownership transfer of RFID tags", *Selected Areas in Cryptography*, p. 276-290, 2005.

[MON 07] MONNERAT J., VAUDENAY S., VUAGNOUX M., "About Machine-Readable Travel Documents", *RFID Security*, 2007.

[NAT 01] NATIONAL INSTITUTE OF STANDARDS AND TECHNOLOGY, Advanced Encryption Standard (FIPS PUB 197), November 2001, http://www.csrc.nist.gov/publications/fips/fips197/fips197.pdf.

[NAT 08] NATIONAL INSTITUTE OF STANDARDS AND TECHNOLOGY, Cryptographic hash Algorithm Competition, http://csrc.nist.gov/groups/ST/hash/sha-3/index.html, 2008.

[NAT 95] NATIONAL INSTITUTE OF STANDARDS AND TECHNOLOGY, FIPS 180-1. Secure Hash Standard, Report, 1995.

[NOH 07] NOHL K., Cryptanalysis of Crypto-1, http://www.cs.virginia.edu./~knsf/pdf/mifare.cryptanalysis.pdf, 2007.

[OHK 05] OHKUBO M., SUZUKI K., KINOSHITA S., "RFID privacy issues and technical challenges", *Commun. ACM*, vol. 48, num. 9, p. 66-71, 2005.

[OKA 92] OKAMOTO T., "Provably secure and practical identification schemes and corresponding signature schemes.", BRICKELL E.F. (ed.), *Advances in Cryptology - CRYPTO '92, 12th Annual International Cryptology Conference, Santa Barbara, USA*, Springer, p. 31-53, August 1992.

[O'N 08] O'NEILL M., "Low-cost SHA-1 Hash function architecture for RFID Tags", *Workshop on RFID Security*, Budapest, 2008.

[ONG 90] ONG H., SCHNORR C.-P., "Fast signature generation with a Fiat Shamir-like scheme", *EUROCRYPT*, p. 432-440, 1990.

[OOR 96] VAN OORSCHOT P.C., WIENER M.J., "On Diffie-Hellman Key Agreement with Short Exponents", *EUROCRYPT*, p. 332-343, 1996.

[ORE 07] OREN Y., SHAMIR A., "Remote Password Extraction from RFID Tags", *IEEE Trans. Computers*, vol. 56, num. 9, p. 1292-1296, 2007.

[PAI 07] PAILLIER P., VERBAUWHEDE I. (eds), *Cryptographic Hardware and Embedded Systems - CHES 2007, 9th International Workshop*, Vienna, Austria, September 10-13, 2007, vol. 4727 of *Lecture Notes in Computer Science*, Springer, 2007.

[PLO 08a] PLOS T., "Susceptibility of UHF RFID tags to electromagnetic analysis", MALKIN T. (ed.), *CT-RSA*, vol. 4964 of *Lecture Notes in Computer Science*, Springer, p. 288-300, 2008.

[PLO 08b] PLOS T., HUTTER M., FELDHOFER M., "Evaluation of side-channel preprocessing techniques on cryptographic-enabled HF and UHF RFID-tag prototypes", DOMINIKUS S. (ed.), *Workshop on RFID Security 2008*, p. 114-127, 2008.

[QUI 00] QUISQUATER J.-J., GUILLOU L., "The new Guillou-Quisquater scheme", *Proceedings of the RSA 2000 Conference*, 2000.

[RAV 01] RAVIKANTH P.S., Physical one-way functions, PhD thesis, 2001, Chair-Stephen A. Benton.

[RIE 05] RIEBACK M.R., CRISPO B., TANENBAUM A.S., "RFID Guardian: A Battery-Powered Mobile Device for RFID Privacy Management", BOYD C., NIETO J.M.G. (eds), *ACISP*, vol. 3574 of *Lecture Notes in Computer Science*, Springer, p. 184-194, 2005.

[RIE 06] RIEBACK M. R., CRISPO B., TANENBAUM A. S., "Is your cat infected with a computer virus?", *PerCom*, IEEE Computer Society, p. 169-179, 2006.

[ROL 08] ROLFES C., POSCHMANN A., LEANDER G., PAAR C., "Ultra-Lightweight Implementations for Smart Devices - Security for 1000 Gate Equivalents", in: GRIMAUD G., STANDAERT F.-X. (eds) Smart Card Research and Advanced Applications, 8th IAP WG 8.8/11.2 International Conference, CARAIS 2008, London, p. 89-103, 2008.

[SAV 07] SAVRY O., PEBAY-PEYROULA F., DEHMAS F., ROBERT G., REVERDY J., "RFID Noisy Reader How to Prevent from Eavesdropping on the Communication?", in: PAILLIER P., VERBAUWHEDE I. (eds) Chyptographic Hardware and Embedded Systems - CHES 2007, 9th International Workshop, Vienna, Austria, p. 334-345, 2007.

[SCH 89] SCHNORR C.-P., "Efficient Identification and Signatures for Smart Cards.", BRASSARD G. (ed.), *Advances in Cryptology - CRYPTO '89, 9th Annual International Cryptology Conference, Santa Barbara*, USA, Springer, p. 239-252, August 1989.

[SHA 08] SHAMIR A., "SQUASH - A New MAC with Provable Security Properties for Highly Constrained Devices Such as RFID Tags", NYBERG K. (ed.), *FSE*, vol. 5086 of *Lecture Notes in Computer Science*, Springer, p. 144-157, 2008.

[SUH 07] SUH G.E., DEVADAS S., "Physical Unclonable Functions for Device Authentication and Secret Key Generation", *DAC*, IEEE, p. 9-14, 2007.

[TUY 06] TUYLS P., BATINA L., "RFID-Tags for Anti-counterfeiting", POINTCHEVAL D. (ed.), *CT-RSA*, vol. 3860 of *Lecture Notes in Computer Science*, Springer, p. 115-131, 2006.

[VAU 07] VAUDENAY S., "On Privacy Models for RFID", *ASIACRYPT*, p. 68-87, 2007.

[WEI 03] WEIS S.A., SARMA S.E., RIVEST R.L., ENGELS D.W., "Security and privacy aspects of low-cost radio frequency identification systems", HUTTER D., MÜLLER G., STEPHAN W., ULLMANN M., (eds), *SPC*, vol. 2802 of *Lecture Notes in Computer Science*, Springer, p. 201-212, 2003.

[WOL 05] WOLKERSTORFER J., "Is elliptic-curve cryptography suitable to to secure RFID tags?", *Workshop on RFID and Lightweight Crypto*, Graz, 2005.

PART FOUR

EPCglobal

Chapter 6

EPCglobal Network

6.1. Introduction

If the history of RFID goes back to World War II, when the allies used it with the FoF (*Friend or Foe*) protocol to identify military aircrafts flying in their airspace, this technology has, however, remained dormant for a long period because of its high cost. Indeed, a much more economical object identification solution was provided by the bar code. This system, which has been largely used for all consumer products since 1970, forms the first effective interface between computers and articles. A reader uses laser beams to capture information on a simple label printed in a set of lines and translates them to a series of numbers: the ID.

This process, widely used today in supermarkets, can track products from manufacture to sale, by passing through storage and shelving systems in stores. However, it suffers from a few limitations. The main one is the constraint of reading, which must be done via an optical contact with the object (with the label precisely). This reading has to be done in the right direction and under the right angle, and without opaque obstacle. This requirement usually requires human intervention for a reading operation, since the information recordable on a label is very limited (in terms of quantity), and is immutable. This major reason for choosing RFID technology, which is moreover faster, for the Internet of Things.

RFID interestingly started to take shape with the creation of the latter concept, when in 1999 the Uniform Code Council (UCC), the regulator of barcodes in the

Chapter written by Dorice NYAMY, Mathieu BOUET, Daniel DE OLIVEIRA CUNHA and Vincent GUYOT.

United States and EAN International (*European Article Numbering*), its European equivalent, founded, in collaboration with Procter & Gamble and Gillette, a laboratory, the Auto-ID Center at MIT (Massachusetts Institute of Technology), USA. The goal of this laboratory sponsored by industries is to perform research in the development of automatic identification technologies. The vision behind these works relates to a physical world represented and controlled by its virtual representation, a kind of intelligent infrastructure allowing us at any time to obtain, through a global network, relevant information on any object through its simple label.

If the initial and the most important motivation of this technology is automating the management of the *Supply Chain*, it should be noted that its application field extends now to many domains such as public transport, health care or automatic payment, etc. This results in the requirement of an important investment to meet the challenges of performance, cost, universality and security posed by a worldwide adoption. From 1999 to 2003, more than one hundred companies and institutions have joined the Auto-ID Center, which finally closed its doors in 2003, with the creation of *EPCglobal*, by the fusion between UCC and EAN International. The largest retailers (Wal-Mart in the United States, Metro in Germany, Tesco in the UK, etc.) and major institutions and industries in various fields, have joined the newly created organization, which is responsible for the global standardization of RFID. Inheriting the work of the Auto-ID Center and associated laboratories (in Australia, England, Switzerland, Japan and China), it started developing its infrastructure: the EPCglobal Network.

The EPCglobal Network refers to the global architecture which allows an object, with values contained in its tag to send the information concerning the product to its server(s). The information passes through various entities and interfaces related to information collection, processing, formatting, resolution, and research sub systems. Finally, the desired information is accesed by the authorized system. In this chapter, we will look at RFID tags, the key of the technology, examine the content (ID), different categories and standards already established. Then we describe the architecture of the EPCglobal Network, including the reader with its interface, the name resolution service (of IDs), the communication language and the service or information server. We also present some security problems, which is the biggest challenge posed by the Internet of Things.

6.2. Tags

6.2.1. *EPC codes*

As mentioned in the introduction to this chapter, the vision underlying the creation of the Internet of Things concept was an intelligent infrastructure that would link objects and related information all over the world. It seems that the primary need is to name, in order to be able to refer effectively, all terrestrial objects – or at least all objects that we want to be part of this network. Thus the EPC code was born: *Electronic Product Code*. This is a serial coding system, where a unique tag is attributed to each object. We will examine various aspects and challenges which are important to consider during the conception of a numbering process – both on the general plan and in this particular case – and we will present the means to respond to them with the EPC code. To facilitate the reading and assimilation, we prefer, whenever possible, to provide end-to-end the needs, solutions and illustrations. The definition of the acronym EPC already contains the word code. Therefore, hereafter, instead of using EPC code, sometimes we will simply say EPC.

The main challenge, which depends on the others, is to list all objects, and identify them in a singular way, without ambiguity. In other terms, an identifier must be associated, every time and everywhere, to one and only one physical item. So we need a code that can satisfy all present and future needs in terms of object identification. A key question arises about the length of the code. The first version of EPC is designed with 96 bits, which corresponds to $2^{96} \approx 8.10^{28}$ possible identifiers, which is more than the total number of items manufactured in the world today. Furthermore, with the development of the idea of the Internet of Things, the identification needs have also evolved. The participation of new application areas such as industry, transport, security systems and many others has resulted in increase of demand for more numbers of objects. We can no longer be limited to individual physical objects. We should be able to correctly identify configurations or durable assemblies of objects such as mechanical systems. Temporary arrangements of objects such as cargoes or pallets can receive virtual EPCs. Even non-objects like services must now be taken into account. As the number of things to be identified becomes difficult to control, it is necessary to ensure that the EPC is always able to satisfy new needs. It is thus important to consider the scalability of the coding scheme.

One of major inherent difficulties in the development of a global standard is to anticipate all possible uses and applications. Not having a perfect vision of the future, it is necessary to envisage an extension method. The EPC design offers the possibility to have several variants. The code includes a header, which allows us to specify a version from which we can deduce its structure. After the first version of the EPC, another type of 64 bits, which is less expensive, was defined. Then another type of 256 bits, a little more exigent, offers more possibilities for object registration. To have more flexibility, the length of the header itself varies, according to the version. However, a problem was raised: to correctly read an EPC, we must first obtain the version that indicates the total length, but how does one read a header where the number of bits is not known? The solution is to allocate version numbers in a way that they are unique for all possible lengths of EPC. The position of the most significant bit with the value 1 determines the EPC length. For 64-bit codes, the header is composed of two bits and the first 1 can be in the first or second position. For 96 bits, it is in the third position in a header of 8 bits. For the 256-bit code, it is in the fifth position in a header of 8 bits as well. The following figure shows the assignments of version numbers for EPC codes.

Today there are seven versions. The value 0000 0000 for the first byte is reserved for a possible extension of the length of the version field, and for future versions of the EPC, the number is limited only by need.

REMARK.– *The version system also allows us to solve the recycling issue. As some organizations may have to follow an object indefinitely, it is important to ensure the longevity of a code. In case of congestion, rather than restarting the numbering and losing old codes, we use a new version.*

Another aspect to consider is the responsibility of the code assignment. Who will allocate the identifiers? How can we function efficiently and ensure that there is never a redundancy? The solution is to distribute this responsibility between EPC managers who are product manufacturers. Each manager manages a part of the EPC namespace which is assigned to him and ensures the uniqueness in his level. The title of EPC manager or the right to assign EPC codes to items is granted after a subscription to EPCglobal, the international organization responsible for the management of EPCs. The subscriber then receives a manufacturer number which will be the common prefix to all object numbers that he creates. For each producer, the number of the object can still be separated into product number and serial number. Thus, a complete EPC consists of four parts:

– the header, described above, provides information on the code structure;

– the producer code, as we have seen, indicates who is responsible for the management of this EPC. The producer code 167 842 659 in decimal is reserved for private uses. Individuals or organizations with private objects which they want to identify can assign the EPCs with this EPC manager value, without having to register with EPCglobal. In order to limit confusion and conflict, it is preferable that only those objects that are not out of the control field of the owner can use this code;

– the product number makes it possible to specify the type or the class of article;

– serial number gives an unique identifier of the considered article in a class or a specific lot.

Figures 6.1 and 6.2 present the partitioning of different EPC versions and a more accurate illustration with a 96-bit code, written in hexadecimal.

REMARK.– *This partitioning also helps us to decentralize services. As we will see later with the ONS, EPCIS and Savants, various entities, companies or servers can be involved in EPC management tasks at different levels.*

Finally, in order to be accepted as much as possible in a global and universal way, the new coding methods should be able to accommodate all old and present

EPC VERSION		VALUE (BIN)	VALUE (HEX)
EPC-64	TYPE I	01	1
	TYPE II	10	2
	TYPE III	11	3
	EXPANSION	NA	NA
EPC-96	TYPE I	0010 0001	21
	EXPANSION	0010 0000	20
EPC-256	TYPE I	0000 1001	09
	TYPE II	0000 1010	0A
	TYPE III	0000 1011	0B
	EXPANSION	0000 1000	08
RESERVED		0000 0000	00

Figure 6.1. *Numbers of actual EPC version*

		VERSION NUMBER	DOMAIN MANAGER	OBJECT CLASS	SERIAL NUMBER
EPC-64	TYPE I	2	21	17	24
	TYPE II	2	15	13	34
	TYPE III	2	26	13	23
EPC-96	TYPE I	8	28	24	36
EPC-256	TYPE I	8	32	56	192
	TYPE II	8	64	56	128
	TYPE III	8	128	56	64

Figure 6.2. *Partitioning of EPC codes*

identification methods and satisfy current industrial coding standards. The EPC code was defined to incorporate the GS1[1] standard. It reuses their structure, always including

---

1 GS1 is a global organism in charge of normalization encoding methods used in supply chain. Its standards include codes for identification of products, containers, places and services (GTIN, SSCC, GLN, GIAI, GRAI, ISBN, etc.). For more details, see [EPC 08a].

ELECTRONIC PRODUCT CODE

01.0000A89.00016F.000169DC0

Header	EPC Manager	Object Class	Serial Number
0–7 bits	8–35 bits	36–59 bits	60–95 bits

Figure 6.3. *Structure of an EPC code in 96 bits*

a Company Prefix, for the company, which corresponds to the EPC manager. The next field indicating the type of product, which can be named in various ways according to the considered code (Trade Item Reference, Asset Type, Location Reference, etc.) is renamed in Object Class in EPC. To complete the latter, we add the version and serial number components if they are not included. Figure 6.4 illustrates this compatibility. The EPC code is compared to the GTIN: Global Trade Item Number, of which the most usual materialization component is the bar code. The EPC is also compatible with all identification supports known today. It can be in the form of bar codes too, it can be registered in RFID tags, or it can simply be manually entered, under the form of characters. Therefore EPC can be used in any application area, irrespective of the entity to be identified, transmission selected, exchange technology or technique.

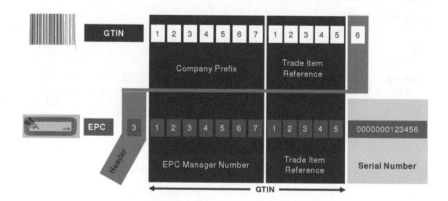

Figure 6.4. *EPC–GTIN compatibility*

6.2.2. *Classes of tags*

There are a variety of tags that can be divided into different types according to the constitution, the memory or calculation capacity, the power source, the transmission techniques and frequencies, the operating distances, the available features, etc. The most popular classification is based on the electrical supply of the tag, so we have:

– passive tags: the tag uses the energy of the radio waves from the reader to all of its operations. A portion of these waves is reflected in a particular way, according to information to send back to the reader, who is the initiator of the communication. This is the most common type of tag. It is sometimes simply referred as RFID tags. Because they have no battery, the main advantage of these tags is their low cost. Furthermore, they are small and have a very long life span as they do not use exhaustible resources. Therefore they are suitable for applications which require massive deployment, but are not very exigent in terms of performance or reliability. These exigences, with the requirement of a small read range, are the characteristics that passive tags lack;

– active tags: different from passive tags, active tags use battery in their operations. Here, we do not really need the reader to initiate the communication, since the tag generates radio waves in a proactive way. These tags are more expensive and have a shorter life span they are available until the battery is discharged. The tag can enter into the field of a reader before activation. It is then "woken-up" by the reader, otherwise it interrupts the power supply to save energy. These tags have the advantage of being more efficient, reliable and have a wider reading field. Therefore, they are suitable for applications that require a lot of resources, for storing (writing on the tag) or calculation (cryptographic functions), without a sensitive price;

– semi-passive tags: like active tags, they have a battery that powers the internal circuit of the tag during communication, which increases the performance, especially the wave scope. However, this battery is not used to generate radio waves. The waves coming from the reader are reflected, like in passive tags. Semi-passive tags combine the advantages of passive and active tags, while reducing the drawbacks of each. They have the reading reliability of active tags and a medium lifetime.

For functioning of transmission frequencies, we can also define: low-frequency tags, which operate in the frequency band 30-300 kHz, high-frequency tags operating between 3 and 30 MHz, the ultra-high-frequency tags with the frequency of communication between 300 MHz and 3 GHz, and microwave tags which operate at over 3 GHz.

Another tag classification can be done according to the constitution and performance of their electronic circuit. The tag can offer only a memory function, and its role is to store an EPC code and return it when asked. Other tags also have a microprocessor to perform operations such as computing cryptographic functions for security.

Another distinction between tags is based on the distance at which they can communicate with a reader. We have close-range tags for which the distance between the tag and the reader must not exceed 1 cm, medium-range tags which can be up to

1 m from the reader, and long-range tags which can operate easily at distances greater than 1 m.

EPCglobal has established its own classification which serves as reference for all actors in its network. It defines four classes of tags, where each class corresponds to a constitution and specific features:

– class 1: ID tags. These are passive tags including an identifier for the object (EPC), an identifier for the tag itself (Tag ID) and a function to kill the tag, which renders it permanently inactive. It may also have a secondary function to put the tag out off service or put it back in service, an access control based on the knowledge of a password and a user memory.

– class 2: high functionality tags. They are also passive, but with more components and features than the tags in Class 1. Here, unlike the previous ones, the user memory and the access control by authentication are required. There is also an extended tag ID.

– class 3: semi-passive tags. As discussed previously, they are passive since they require the presence of a reader to initiate communication, but they have a proper energy source. They can also include sensors, which can record the collected data.

– class 4: active tags. Like the Class 2 tags, they have an EPC, an extended tag ID and an access control with authentication. They have a battery as the power source, and communications are established via an autonomous transmitter which is able to initiate exchanges with a reader or another tag. They can also have a user memory and sensors that can record the collected data.

There are a multitude of tags with various features which are suitable for specific applications. Knowing that these parameters are sometimes dependent (the communication distance depends on the magnetic field strength) and a particular advantage can generate some specific disadvantages (a tag with microprocessor is more expensive than a simple tag with memory), it is necessary for each application to select and make choices based on the target objectives.

6.2.3. Standards of tags

The tag is the terminal element of the infrastructure, and is mainly designed to carry a code, where the code is communicated to a reader by radio transmission, when necessary. In this point, contactless smart cards can be considered as RFID tags, especially when they are used for identification. Although they are nowadays more powerful, this difference tends to disappear with technological evolutions, especially for new "smart" tags. This section presents the norms and standards of tags and extends up to smart cards. We begin with the utility of the standards and then present some of the most common ones.

The world of smart cards and RFID tags includes a multitude of manufacturers. For a smooth running of a global network as defined above, it is necessary that different actors such as industries, business or individuals can easily communicate, independent of the material. A good interoperability between systems and RFID devices requires prior agreements on the structure and signification of exchanged data, as well as on the mechanisms used for communication. Beside the need to communicate with a diversity of material, there is another requirement related to the non-interfering with other equipments. In order to reduce electromagnetic wave interferences, it is important to consider the frequencies that are already being used, in a global way or in limited geographic areas, and to use appropriate devices or frequencies. It would be interesting to have frequencies or frequency ranges determined locally – or globally if possible – and dedicate them to RFID. Another adequacy of this standardization is related to public health and environmental protection. Indeed, if a large-scale deployment of RFID promises significant benefits in everyday life, it is necessary to consider the impact of radio wave radiation on people and their environment. The adoption of standards that are inoffensive to humans in particular and are respectful to nature in a general sense, established under investigations and verifications by competent working groups, is a necessity for ensuring humanity consciousness of these values.

Given these constraints, specifications have been defined to standardize the various aspects of the tag operation. These standards determine, in particular, transmission power frequencies, the type and the duration of communication cycle. We present briefly some of them here:

– Standard ISO/IEC: ISO (International Standards Organization) and IEC (International Electrotechnical Commission) are two organizations which form a worldwide technical standardization body. Their members are national organizations who participate in the establishment of standards by sending representatives to the committees that are established for various technical activity areas and issues. These committees have defined standards in the radio-frequency communication field, including contactless smart cards and RFID tags:

- ISO/IEC 14443: commonly known as ISO 14443, it is a standard for identification contactless integrated circuit cards, called proximity cards and operates at 13.56 MHz. This standard defines physical characteristics of the card and communication protocols with the reader. It consists of four parts:

Part 1. Physical Characteristics
This standard defines the size of the card, the quality or the type of surface (for printing) and the behavior in some environmental conditions such as X or UV radiation, temperature or ambient electromagnetic fields. It even specifies the resistance to mechanical stresses such as folding or twisting. These requirements must be fulfilled at the card level, and this depends on the card manufacture.

Part 2. Radio frequency power and signal interface
This standard describes the power transfer and the communication between the card and the reader. The chip, which does not require a battery, receives energy from waves emitted by the reader. The transmission frequency is fixed at 13.56 MHz. Two types of communication interfaces are defined, type A and type B.

Their differences lie in the modulation of the magnetic field, the data encoding format and the anti-collision technique.

Part 3. Initialization and anti-collision

This part of the standard describes the selection of a tag entering into the field of a reader which "speaks" first, the formats and the rhythm of anti-collision commands (requests and responses). It also defines the anti-collision methods. As mentioned above, these methods depend on the communication type. For the type A, it is a binary search method with the unique tag identifier as reference. The type B uses the *Slotted Aloha*[2] method.

Part 4. Transmission protocols

This part is optional, as the first three parts are considered sufficient to meet the standard. It specifies a half duplex transmission protocol. This standard also defines the information exchange which is independent of lower layers. For that, high-level data transmission protocols are defined for the types A and B. These protocols determine the data encapsulation into blocks and the error management and support.

- ISO/IEC 18000 or ISO 18000. This is a series of standards defined for RFIDs in view of identification objects. They define data transport protocols without dealing with the content or structure of the data or the physical implementation of tags. They have the peculiarity, if possible, to dedicate the same protocols to devices using different frequencies, thus reducing interoperability problems. There exists a description of the RFID system architecture, which defines a set of common parameters, valid for all frequencies, and a set of specifications, specifying specific communication parameters for each of the legal frequencies or frequency ranges. Therefore, ISO 18000 is divided into seven sections as outlined below:

Part 1: the goal of this standard is to describe a reference architecture for object management. It also establishes generic parameters, determined in the other standards of the series. Indeed, subsequent parts provide specifications for a particular frequency or frequency range. It is a permissive standard, which supports different RFID implementations without making any distinction of their related technical qualities.

Part 2: this section specifies the parameters for communication between tags and readers at frequencies below 135 kHz. It defines the communication protocol, commands, and singulation and anti-collision methods. It also determines the physical layer used for the communication between the reader and the tag for two main types, A (full duplex, continuously powered by energy from the reader operating at 125 kHz) or B (half duplex, powered by the reader except during the tag-reader transmission, it operates at 134.2 kHz).

---

2 *Slotted Aloha*: anti-collision used for the tag singularization, in which time is discretized in constant intervals, two packages are in collision only if they are ready to emit in the same interval.

Part 3: this section focuses on the communication for tags at 13.56 MHz. It defines communication parameters for three modes based on supported transmission rates. For example the mode 1 can have a throughput of 26 kilobits/s in both directions, while the mode 2, called high rate, can go up to 105 kilobits/s from tag to reader and 423 kilobits/s in the reverse direction.

Part 4: this standard considers RFID devices operating at 2.45 GHz in the ISM (Industrial, Scientific and Medical) domain. It supports implementations for wireless devices, in information system side, which allows applications to operate at a range of more than one meter. It defines two communication modes, depending on whether the tag is passive or active.

Part 5: this part defines the physical layer, communication protocols and collision processing for tags operating at 5.8 GHz. It has been abandoned due to lack of global interest.

Part 6: This standard defines the parameters for communications at 860-960 MHz. Like all others, it describes the physical interactions between tag and reader, protocols, communication commands and collision handling. It considers three types of tags, distinct by encoding, transmission rate and collision management. In particular, the type C is identical to the specification "UHF Gen 2" of EPCglobal.

Part 7: This standard defines the parameters for the communication of RFID equipment at 433 MHz. The considered devices have advanced features such as the writing on the tag, the tag selection by address groups or the error detection. Like Section 4, it considers the passive and active tags whose range can reach up to 1 m.

– EAN/UCC standards: The former *EAN International* and *Uniform Code Council* have jointly developed a set of standards for the supply chain. These standards initially dealt with exchanged data, defining different product codes (GTIN, GLN, SSCC, etc.) and possible representations or symbols (bar codes, data matrix, etc.) and have been extended to the RFID technology itself. The GTAG (Global Tag) project, launched in 2000 for the definition of automatic identification standards by radio frequency for the supply chain, extends to the communication interface between tags and readers. They are application norms, related to the use of products, contrary to technical standards dedicated to manufacturers as previously presented. Today a standard has been ratified by EPCglobal for ultra high frequency tags, the standard "Class-1 Generation-2 UHF RFID", which reuses the ISO 18000-6 and handles the communication for tags at 860 - 960 MHz. Another standard under development is for high-frequency tags. Before emergence of these standards, EPCglobal specifications have existed, given from the work of the ex-Auto-ID Center at MIT and used as normative references. They concern UHF tags at 900 MHz or 860-930 MHz and HF tags at 13.56 MHz. All these standards define communication protocols and standards between tag and reader, tag singulation methods, (when several tags are in the reader field) and ways to reduce interferences.

There are other standards such as ISO/IEC 15693 for contactless identification cards called vicinity cards, ISO/IEC 10536 for Identification close-coupled cards, etc. Furthermore, in parallel with global standards such as ISO/IEC or EAN/UCC, there exist, at national and regional levels, laws and special regulations governing the use of RFID in the concerned territories. They often result in recommendations and decisions which should be used as standards. There is for example the case in Europe with recommendations established by ERC, European Radio Communications Committee, for short range radio devices – which are naturally a part of RFID tags. They define the terms of use concerning frequencies, power, bandwidth, cycle time and even applications.

The standards aim at facilitating the interoperability of equipment despite the differences between manufacturers or regions. But as we can see, in the world of smart cards and RFID tags, a multitude of standards exist and it is still an obstacle for an adoption of global tools with the same characteristics. EPCglobal, responsible for setting global standards of RFID, participates in this harmonization. However it is not always easy because in some countries or regions, the frequencies proposed in the standards are already in use or prohibited (for example in France, the army uses the frequencies 865-868 MHz for its tactical relays which are selected by European regulations for RFID).

6.3. EPCglobal architecture

6.3.1. *Reader protocol*

The reader's protocol allows tag readers to interact with application software according to the EPCglobal standard. The terms "tag reader" or "reader" include RFID tag readers, irrespective of the used radio frequencies. They also characterize the readers of other types of tags, such as bar codes. Tag readers, despite their name, should also be able to write data into tags.

The reader protocol specifies the interactions between a device capable of reading (and in some cases writing in) tags, and an application software. The two parts are called reader and host. An example of a host is a middleware or an application compatible with EPC. Nevertheless, the reader protocol does not require the use of a middleware or a particular application.

The role of the reader protocol is to isolate the host that does not know the details of the interaction between the reader and tags. Readers can use different protocols to communicate with tags (not only radio tags, a reader may also be able to read barcodes visually. However, only the reader protocol is used during the communication between the reader and the host.

Reader Protocol (RP) is an interface standard specifying the interactions between a device capable of reading/writing in tags and an application software. The goal was to define an open and extensible interface which reader vendors can adopt to support most of the operations in a standard way.

Features make it possible to read, write, disable tags and access the user memory. They also enable identity information, configuration of commands, information report and options, and asynchronous notifications.

6.3.1.1. *Different protocol layers*

As shown in Figure 6.5, the reader protocol is specified in three distinct layers:

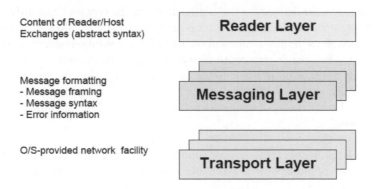

Figure 6.5. *Reader protocol*

– reader layer: this layer specifies the content and abstract syntax of messages exchanged between the reader and the host. This is the core of the reader protocol since it defines the operations that the reader can perform;

– messaging layer: this layer specifies how messages defined in the reader layer are formatted, encapsulated, transformed, and transported in the network;

– transport layer: this layer corresponds to network communication possibilities provided by the operating system (or its equivalent).

The reader protocol specification provides many possible implementations in the message and transport layers. Each implementation is known as MTB (*Messaging Transport Binding*). Different MTBs provide different kinds of transport such as TCP/IP by Bluetooth or a traditional serial connection. Different MTBs can also provide different ways to establish connections (the reader contacts the host, the host contacts the reader, etc.). They can provide initialization messages required to establish the synchronization, or providing information of the configuration.

6.3.1.2. *Message channels*

The interface between a host and a reader is defined in message channels; each one represents an independent communication between the reader and the host. Two types of message channels are defined:

– control channel: it transports requests from the host to the reader, and responds to the requests transported from the reader to the host. All messages exchanged in the control channel follow the request /respond scheme;

– notification channel: the notification channel transports messages asynchronously sent by the reader to the host. The messages in the notification channel are only sent by the reader. The notification channel is mainly used to enable an operating mode in which the reader delivers information read on tags to the host without any requirement.

Two channels are defined, so that notifications can be sent to a host different from the one that sent the commands. This task is performed by the commands of the control channel which allow the host of the control channel to specify a secondary host to which the reader will be able to connect to deliver notifications. In some cases, the reader can allow the notification issuance to the same host which sent the commands in the control channel, with the traffic multiplexed in the same transport connection. Drives connected to hosts (via serial connections for example) often require this notification delivery mode.

The reader protocol can assign one or more control channels to a single reader. In some cases, one or more notification channels of a reader can be used with one or more hosts.

6.3.1.3. *Link to the reader layer*

The reader protocol provides a unique way for hosts to access and control compatible readers from different manufacturers. There are all kinds of readers which provide different features. A basic reader will provide information on tags located in the radio area, while a more advanced reader will be able to, for example, filter in a sophisticated way.

The reader protocol defines a set of features which are implemented in readers and provides a standard way to access and control these features when they are present. The reader protocol does not require that all readers implement all these features. But if it is the case, it allows hosts to access them in a standardized way.

6.3.2. *Application Level Events (ALE) interface*

The purpose of the ALE interface is to provide filtered and aggregated data from different sources to clients. The idea is to include a processing level which analyzes data from different EPC data sources in order to identify important events for applications. In addition, this interface adds an isolation layer among end-user applications, i.e. clients, and the physical layer, developed and maintained by other entities. This isolation layer, shown in Figure 6.6, increases the system flexibility and reduces the development cost of the component of the architecture for either clients or managers of physical infrastructure.

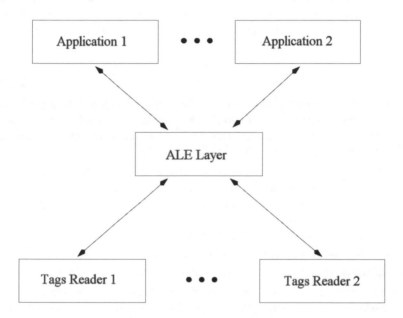

Figure 6.6. *Organization of layers*

The functions supported by the ALE layer are:

– reception of EPC information from multiple readers;

– accumulation, filtering, counting and grouping of data in order to remove redundancies, reduce the information volume and facilitate the application execution;

– providing different types of reports on collected data.

6.3.2.1. *Evolution of the Savant middleware*

Initially, these functions were grouped in a particular component of the EPC architecture, called Savant. This terminology was used to refer, in a general way, to any software located between RFID readers and final applications. In a more specific manner, the term is related to the particular structure proposed by the Auto-ID Center at Massachusetts Institute of Technology (MIT) for this software [OAT 02, CLA 03].

In order to avoid ambiguity, EPCglobal has abandoned this terminology [CPR 05] in the most recent specification whose main objective is the definition of different features. This specification includes a formal processing model, an API (*Application Programming Interface*) in UML and binding points between the API and a SOAP (*Simple Object Access Protocol*) protocol compatible with the WS-I (*Web Services Interoperability*) specification. Contrary to earlier specifications of Savant, the latter one focuses on an external interface, without considering the implementation or internal interfaces of each software part. The approach adopted by EPCglobal admits several implementation possibilities having a common external interface.

6.3.2.2. *Specification flexibility*

As noted before, the objective of the ALE layer is to provide an isolation between the application and physical layers. To obtain this isolation and to allow the development independent from client application and reader infrastructure, the interface as described in the EPCglobal specification [EPC 05] has three characteristics:

1) the proposed specification provides methods for clients to specify the searched EPC data in a high-level declarative way without imposing implementation constraints. The interface is developed, so as to allow the possibility of implementing any strategy adapted to meet client's request. For example, this strategy can be optimized vis-à-vis the desired performance and specific capabilities of readers;

2) the specification provides a standard format for report generation using EPC data obtained independently of the locations where they were collected;

3) the specification is an abstraction of EPC data sources in high-level *logical readers* which are in general related to the location. This abstraction can hide the physical readers used to collect EPC data in a given logical location. In this way, the type and quantity of used readers as well as aspects related to the communication between readers and tags become transparent to clients.

Figure 6.7 illustrates the advantage of using high-level abstractions related to the location. Indeed, using this figure, one can verify two distinct situations. In the first case, the first logical location, *Location 1*, has only one reader, L1, and the second logical location, *Location 2*, has two readers, L2 and L3. If the client does not need low-level information on the used readers, the complexity to address the two locations is the same. The fact that the Location 2 has two readers becomes transparent for client. In a second scenario, the person responsible for the handling of the system at the physical level decides to insert a new reader L4 in Location 1 to improve signal quality. This insertion produces a change in the environment (as indicated by the arrow). The use of the ALE interface makes the change transparent to the client who continues to address to Location 1 in the same way as he did before the change. This independence between the application and physical layers, obtained by the ALE interface, increases the architecture flexibility and reduces handling costs.

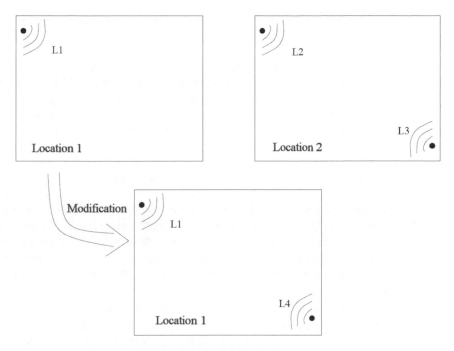

Figure 6.7. *Logical locations*

6.3.2.3. *Role of the interface in the EPCglobal architecture*

Many reading operations will be generated inside the production chain of systems which implement EPC technology. Moreover, it is highly likely that an important part of these readings represent "noise" for clients of the system. The EPCglobal architecture is designed, so that reading, filtering and counting are made as low as possible in the architecture.

The specification of the ALE interface facilitates this goal by providing a flexible interface to a standard set of accumulation, filtering and counting operations which produce "reports" in response to "requests" from clients. From their side, clients are responsible for the interpretations and decision making, related to the content of these reports.

The ALE interface is built around client requests and reports generated in response. These reports can be directly sent to the entity making the request or to a third party appointed by the entity responsible of the request. The system provides two types of requests:

– immediate requests. Reports are generated one time in direct response to a request;

– recurrent requests. The reports are sent each time an event of interest is detected or at a regular interval.

Logically, these features are provided in a layer between client applications and physical readers (Figure 6.6). However, it should be noted that the ALE interface specification does not specify where this processing is done in practice. The processing can be based on a middleware entity completely independent or performed in a reader with sufficient resources. In all cases, the ALE interface functions as a reference point for the interface with client. In this way, it gives more liberty to system architects, allowing them to exploit the capacity of readers to the maximum, while ensuring that clients have a black box of which only the interfaces must be known.

In several application scenarios, the ALE client is a software that incorporates the *EPC Information Service* (EPCIS), or any other similar processing software. As EPCIS also processes high-level events, it is important to distinguish between these two components of the EPCglobal architecture. In addition, since EPCIS is oriented to data processing for business applications, the processing type is different. The main differences are:

– the ALE interface is oriented only to the real-time processing of EPC data without the need to register data in a persistent way. Of course, the client can decide to register them in order to increase the system robustness, but this feature is optional. However, business applications typically use historical data, which are naturally persistent;

– the communication of events through the ALE interface uses a simple syntax of type *what*, *where* and *when* without a business semantic. On the contrary, data in EPCIS level generally have a business semantic. Thus, an event in the ALE level can simply provide information related to reading of some tags at a given time and given location, when, at EPCIS level, the information can correspond to the addition of objects associated to tags in the stock.

The ALE layer handles the flow of data acquisition and filtering to identify significant events. Obtaining a subset of significant events is the starting point for the application of business logic. Application layers of a company handle the process at this level and the recording of events that can relate to other information processing activities in the company.

6.3.3. *Object Name Service (ONS)*

The Object Name Service (ONS) [EPC 08b] is used to locate metadata which are authorized on a given Electronic Product Code (EPC) and the associated services. It uses the existing Domain Name System (DNS) [MOC 87a, MOC 87b], which is used on the Internet to establish the correspondence between an IP address and domain name, to collect information. Its role is to translate an EPC into one or several *Internet Uniform Reference Locators* (URLs) in order to seek more information. In general, these URLs identify an EPC Information Service (EPCIS). However, the ONS can also directly associate an EPC to a website or other resources accessible via the Internet.

6.3.3.1. *DNS*

DNS is a hierarchical naming system for computers, services, or any other resources connected to the Internet. It associates various information to domain names assigned to participants. It translates domain names which are strings of characters easily memorable and comprehensible to humans, into binary identifiers of network devices. These binary identifiers, typically IP addresses, are then used by other protocols and services to locate or contact machines. For example, the URL *www.example.fr* is translated into *132.227.110.107*.

The DNS namespace is organized as a tree. It is divided into zones that start at the root. Each DNS zone is a collection of nodes placed under the authority of an authoritative name server. These servers record the changes occurring in their areas of authority to avoid continuous accessing and updating the central registry. Finally, the administrative responsibility for a zone can be divided to create additional zones. The authority is then delegated, under the form of sub-domain, on a part of the old namespace in the area. The top of the hierarchy consists of root servers. The requests concerning the names of TLDs (*top-level domains*) are sent to them.

A domain name consists of two parts, called labels, which are conventionally separated by dots, for example: *example.fr*. The label in the right, *fr*, corresponds to the TLD, i.e. to the area directly under the root. Each label on the left is a sub-domain of the domain above it. For example, *example.fr* is a sub-domain of the domain *fr*.

On the client side, the DNS requests aim to recover the complete resolution (translation) of the research resource, for example, the resolution of a domain name in the corresponding IP address. The DNS queries can be separated into two types: recursive or iterative. When a DNS server receives an iterative request, it returns a partial response or an error. From this information, the client contacts another DNS server which returns a partial response or an error. The client runs through the tree until the DNS server containing required information is reached. On the contrary, when a DNS server receives a recursive request, if it does not have all the required information, it will contact another DNS server in the hierarchy. This server will follow the same process, and so on, until the good server is contacted. At that time, the response follows the inverse path and the DNS server contacted by the client sends required information or an error if they are not found. This mode is optional for DNS servers.

6.3.3.2. *Utilization of DNS by ONS*

In order to use DNS to find the information corresponding to an object, the EPC of the object must be converted into a format that can be understood by the DNS, i.e. the

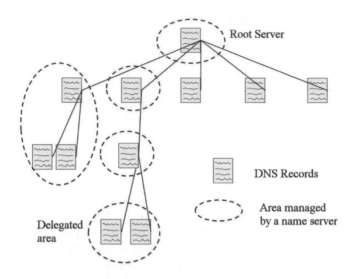

Figure 6.8. *Structure of DNS*

domain names delimited by points. The EPC resolution mechanism requires that the handled EPC are as per the URL form, as defined in the standard [BER 05].

On the same principle as DNS, ONS consists of local servers. The terminology distinguishes them into two types. The local ONS servers process requests concerning the EPCs that are under the control of the company managing them. The root ONS server, while receiving a request, sends a list of available services which correspond to the required EPC. They contain only network addresses of services that contain data. Figure 6.9 describes a typical ONS request:

1) a sequence of 64 bits representing an EPC is read from an RFID tag. For example: *10 000 00000000000000 0000010000000000011000 000010001001100011001 0000*;

2) the reader sends this sequence of bits to a local server. For example: *10 000 00000000000000 0000010000000000011000 000010001001100011001 0000*;

3) the local server converts the bit sequence in URL according to the standard [BER 05]. For example: *urn:epc:id:sgtin:0614141.000024.400*;

4) the local server sends the URI to the ONS. For example: *urn:epc:id:sgtin: 0614141.000024.400*;

5) the ONS resolver converts the URI domain name and sends a DNS request for that domain. For example: *000024.0614141.sgtin.id.onsepc.fr*;

6) DNS returns a set of answers that contain URLs pointing to one or more services (for example, an EPCIS server);

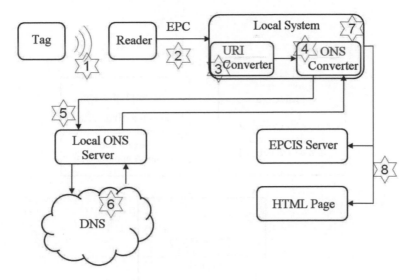

Figure 6.9. *An ONS request*

7) the ONS solver extracts the URLs of DNS responses and forward them to the local server. For example: *http://epc-is.exemple.fr/epc-phare.xml*;

8) the local server contacts the EPCIS server specified in the URL for EPC.

6.3.4. *Physical Mark-up Language (PML)*

Physical Mark-up Language (PML) describes how information is transferred in the EPC network. It provides a collection of shared and standardized syntaxes to represent and distribute information inside the EPC network. It consists of two sets of syntax: *PML Core* and *Savant Extension* (Figure 6.10).

The Savant extension is used for communications between Savant and business applications.

The PML Core defines the standard format for data transfer. It should be understood by all nodes of the EPC network, i.e. by ONS, Savant and EPCIS, to ensure data exchange and facilitate their configuration. It should also be easily legible by humans. PML language is therefore based on the *Extensible Markup Language* (XML) standardized by W3C [BRA 06]. Codes are used to format data before they are sent. For example: *<lment></lment>*. Three types of components can be described: sensors, observations and observables. The sensors are, in a general way, devices capable of performing physical measurements, such as temperature, or logical measurements such

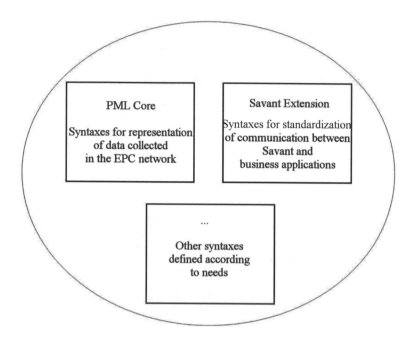

Figure 6.10. *The PML is a collection of vocabularies that represent information related to objects in the EPC network*

as an identity. This category includes RFID readers, barcode readers, sensors and very-small-capacity devices. The observations represent the measurements made by sensors. They associate current data in sensors. Finally, observables are physical properties or entities observed by sensors.

An XML example for a tag element:

```
<pmlcore:Sensor>
    <pmluid:ID>urn:epc:1:4.16.36</pmluid:ID>
    <pmlcore:Observation>
        <pmlcore:DateTime>2006-11-06T13:04:34-06:00</pmlcore:DateTime>
        <pmlcore:Tag>
            <pmluid:ID>urn:epc:1:2.24.400</pmluid:ID>
        </pmlcore:Tag>
        <pmlcore:Tag>
            <pmluid:ID>urn:epc:1:2.24.401</pmluid:ID>
        </pmlcore:Tag>
    </pmlcore:Observation>
</pmlcore:Sensor>
```

6.3.5. *EPC Information Service interface*

EPC Information Service is the specification of a standard interface to access information of the EPC system. Since an EPC (Electronic Product Code) attributes a unique serial number to each object, each object can be individually tracked while collecting information of each object in real time. The EPC Information Services provide means for all supply chain partners to effectively share and exchange information. Indeed, having a standard interface allows different actors to use the same functions to access data through the supply chain. The results in the reduction of the integration duration between partners when everyone uses the same interface, even if these partners record information in different databases. Figure 6.11 shows the EPC Information Service.

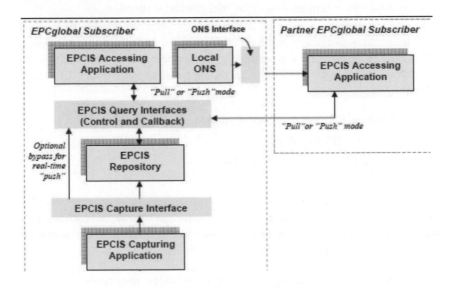

Figure 6.11. *EPC Information Service*

EPCIS is mainly concerned with sharing of data at the serial level. EPC Information Service has a distributed architecture, and is a technical specification for a communication interface. This is not a service centrally managed by EPCglobal.

Data related to EPC are divided into two categories:

– the temporal data collected during the lifecycle of an object, such as measurements (data from sensors such as temperature history of a product);

– data attributed in a quasi static way in the serial level, without need to update, as per the production date or the expiration date.

6.3.5.1. *Specification*

The EPCIS working group of the EPCglobal Software Action Group writes the technical specification of the EPC Information Service. It is composed of end user representatives and technology solutions providers. End users interested in conditions and technology solution providers focus on technical implications.

The EPCIS interface specification is specifically intended to avoid favoring any particular choice of implemented components:
- database (relational, object-oriented or document-oriented);
- operating system (Windows, Mac OS, Linux, etc.);
- programming language (Java, C#, etc.);
- integration with information systems of a particular supplier.

EPC Information Service is based on the top layer of EPC network technologies. EPCIS is the first layer where the business logic can be mixed with events read from RFID readers. All the layers under EPCIS manage data triplets: reader, EPC tag and timestamping.

The EPC Information Service specifies standard interfaces for:
- interrogate (get EPCIS data);
- events capture (record data in EPCIS).

6.3.5.2. *Implementation*

To implement an EPC Information Service, one can choose to host own EPCIS interface coupled with its existing databases for serial level data, or can also register to a technology solution provider hosting an EPCIS service.

For a manufacturer, trading partners will probably be able to find their EPCIS by using Object Name Service (ONS), searching the EPC code of its products. For a distributor or a retailer, it is uncertain whether the ONS will also point on EPC Information Services. Indeed, ONS uses the DNS technology, which is sufficiently robust and mature for the Internet, but is not adapted to real-time. Another obstacle in the use of DNS technology is its lack of security, whereas data streams are composed of sensitive data.

6.3.6. *Security*

EPCglobal Network is a global network of traceability, in the service of industrial, commercial, logistics providers and even consumers to monitor the progress and status of goods from production to retail. If it has certain advantages, for the supply chain management with object tracking, it is important to note that an indelicate massive deployment could cause serious damage to security, because of the possibility of information leakage at different levels of its architecture. In this section we examine dangers or security flaws of EPCglobal Network, and some implemented or possible methods to remedy them.

The structure of the EPCglobal network as defined above includes several entities (tag, reader, middleware, EPCIS, ONS, etc.). We will analyze the security level of each unit. There are also security problems specific to the interfaces between different pieces of the architecture. The problems are in general comparable to the ones of the related entities. We will not consider the separation between the architecture bricks and interfaces except the interface tag-reader with relatively considerable problems.

6.3.6.1. *Tags*

The first threat for the tag is that it can be torn from its article, in order to, for example, pass a reader in the supermarket without being noticed, or to deceive an article by pasting another tag on the object. To prevent this, product and tag manufacturers should ensure that they cannot be removed without obvious damage to the object or to the tag itself.

Thieves can easily clone the tags in cases where extraction and reuse of the tag is difficult. One needs only to read its EPC and make another tag with the same code. The solution is that the tag should transmit its data only to safe and authorized readers. An authentication of the reader to the tag is therefore required. It uses cryptography, with passwords, encryption and hashing functions, which ensure that only an authorized reader can access and/or decrypt the data transmitted by the tag. Similarly, it is necessary that the tag can be authenticated to the reader, for clonning detections.

Some tags are rewritable, and in the absence of effective measures malicious readers can illegally modify its content. Access in writing to a tag should be possible only for an authorized reader. To ensure this restriction, an access control, based on, for example a password, should be set up in the tag. This is the case for UHF Class 1 Gen 2 tags of EPCglobal, which have a password of 32 bits to protect the access to the memory of the tag.

6.3.6.2. *Reader-tag communication interface*

The communication between reader and tag can be eavesdropped by a malicious reader. Eaves dropping could be used by the malicious reader for replay or man-in-the-middle attacks in which the malicious reader is positioned between two communicating devices. Facing these two devices, the reader masquerades as the other device. To counter such attacks, previously proposed solutions can be used, such as encryption and mutual authentication, thus preventing any indiscreet eavesdropper from extracting information. For a relay, one may also render collected information useless by changing the value sent by the tag in each interrogation, according to an algorithm and parameters controlled by tag and reader. For the UHF Class 1 Gen 2 tag of EPCglobal, random numbers generated by the tag in each session are used to encrypt (by an X-OR operation) the password sent by the reader. This approach is not safe enough since the considered random numbers are sent in clear.

Another serious threat on the interface between the reader and the tag is the DoS (Deny of Service) attack. The radio frequency channel could be invaded by signals generated by a malicious reader in order to harm all communications. There is no absolute solution for this danger, we could simply install a device capable of blocking waves coming from the outside of an environment to protect.

6.3.6.3. *Reader*

A malicious reader placed, for example, in a warehouse by a competitor, like a compromised reader, is a serious threat. In the same way, any reader in the neighbourhood should be authenticated before any transaction to ensure protection. This can be done for example by certificate verification. A permanent surveillance is required, for ensuring that readers do not keep information collected from tags, and that all their communication ports are used only by the company. The physical access to the reader and its configuration must be checked and authorized to specific and known persons only.

Like the interface between the tag and reader, the one between the reader and the middleware can be attacked by eavesdropping, replay, or man-in-the-middle attacks. To avoid attacks, we need mutual authentication between the reader and the middleware. A secure communication channel can be established between them, for example using SSL-TLS, EAP-TLS, X.509 certificates, public keys, signatures, etc.

6.3.6.4. *Middleware*

The part of the EPCglobal architecture located between the reader and the server holding data is often called middleware. It can also be seen as an application server and thus suffers from all the threats related to the server function, including intrusion, DoS attacks, viruses, sabotage, etc. At this level, authentication, authorization and access control are dominent. It is also necessary to have intrusion detection systems and logging, antivirus and firewalls, etc.

6.3.6.5. *EPCIS*

EPCIS, "database" containing information about objects, is no longer safe from attacks. Its interface with the middleware must be protected like the reader interface from eavesdropping, replays, man-in-the-middle attacks, etc. The same measures akin to the ones taken to protect the middleware, are applicable here (authentication, access control, etc.). It is also important, as for any server, to have backup copies against possible destructions or "mirrors" to reduce dangers of a DoS attack.

6.3.6.6. *ONS*

ONS, based on DNS, can suffer from the same attacks as the latter, such as the corruption of system files or cache, false IP addresses, data interception, etc. Countering these threats thus requires a good system administration (file system backup, access control with a proper permission allocation, etc.), an authentication of the origin of DNS data (certificates, signatures, etc.) and systems such as firewalls or intrusion detection.

6.3.6.7. *Authentication service of subscribers*

Information about objects are produced, maintained and kept by EPCglobal Subscribers. They are industrial, commercial companies and logistic service providers subscribed to EPCglobal. They should be able to provide data in the network (only Gillette can generate information on Gillette brand items). Similarly, only authorized persons, firms or institutions can access them. Therefore, an authentication of subscribers is needed to receive and transmit information without anarchy. This is the reason why we need this authentication service. It requires no prior knowledge or arrangement between two subscribers. Like any server, the authentication subscriber must also be protected from intrusions, viruses, DoS, etc.

6.4. Conclusion

The global tracking network EPCglobal was born to cater for the needs of industries, logistic providers and retail actors, for tracking objects in real time, from production firms to stores, and finally to consumers. Registration numbers (EPCs) assigned to objects are embedded in RFID tags where simple readings make it easier to identify objects and obtain other information about them. Between the reader and the data server (EPCIS), the architecture consists of Object Name Server (ONS) and data processing interfaces (formatting, filtering, aggregation, etc.).

However RFID, associated with EPCglobal in its wide field of applications and fabulous benefits in everyday life, also presents serious threats to security at any level to individuals, companies or institutions. In order to take advantage of the contributions of such technology, without suffering potential negative effects, research works have led to a lot of solutions. These solutions are applicable to different architecture levels of this network and allow protection of the identity tag during communications, operation controls on tag (deactivation, writing, etc.), provision of data access surveillances and countermeasures against attacks aimed at harming the system (viruses, denial of service, etc.). However, there is no universal solution, each has its own advantages and disadvantages (regarding the performance or cost for example) and corresponds to a specific application. Again, we should make a choice in each situation, depending on the constraints to be respected and available means.

Furthermore, it should be noted that if the structure of the EPCglobal network is already well defined and components are correctly identified, the establishment of standards is still ongoing and is a long process, which can delay the worldwide adoption of this technology. In addition, the aim of the vast majority of businesses and different supposed actors of the network is to inform and to encourage all industries to join in.

6.5. Bibliography

[ALB 07] ALBERGANTI M., *Sous l'oeil des puces : La RFID et la démocratie*, 2007.

[BER 05] BERNERS-LEE T., FIELDING R., MASINTER L., *Uniform Resource Identifier (URI): Generic Syntax*, 2005.

[BRA 06] BRAY T., PAOLI J., SPERBERG-MCQUEEN C.M., MALER E., Extensible Markup Language (XML) 1.0 (Fourth Edition), Report, W3C, August 2006.

[BRO 01a] BROCK D.L., *The Compact Electronic Product Code: A 64-bit Representation of the Electronic Product Code*, November 2001.

[BRO 01b] BROCK D.L., *The Electronic Product Code (EPC): A Naming Scheme for Physical Objects*, January 2001.

[BRO 02] BROCK D.L., *The Virtual Electronic Product Code*, February 2002.

[CLA 03] CLARK S., TRAUB K., ANARKAT D., OSINSKI T., SHEK E., RAMACHANDRAN S., KERTH R., WENG J., TRACEY B., Auto-ID Savant Specification 1.0, Auto-ID Center Software Action Group Working Draft WD-savant-1_0-20031014, October 2003.

[ENG 03a] ENGELS D.W., EPC-256: The 256-bit Electronic Product Code Representation, Report, February 2003.

[ENG 03b] ENGELS D.W., The Use of the Electronic Product Code, February 2003.

[EPC 05] EPCGLOBAL, The Application Level Events (ALE) Specification - Version 1.0, EPCglobal Ratified Specification, September 2005.

[EPC 08a] EPCGLOBAL, EPC Tag Data Standards Version 1.4, EPCglobal Tag Data and Translation Standard Work Group, June 2008.

[EPC 08b] EPCGLOBAL, "EPCglobal Object Name Service (ONS) 1.0.1", Ratified Standard Specification with Approved, Fixed Errata, May 2008.

[GAR 06] GARFINKEL S., ROSENBERG B., *RFID: Applications, Security and Privacy*, 2006.

[KON 06] KONIDALA D.M., KIM W.-S., KIM K., *Security Assessment of EPCglobal Architecture Framework*, 2006.

[MOC 87a] MOCKAPETRIS P.V., *Domain Names - Concepts and Facilities*, 1987.

[MOC 87b] MOCKAPETRIS P.V., *Domain Names - Implementation and Specification*, 1987.

[OAT 02] OAT SYSTEMS, MIT AUTO-ID CENTER, The Savant Version 0.1 (Alpha), MIT Auto-ID Center Technical Manual MIT-AUTO-AUTOID-TM-003, February 2002.

[SAR 00] SARMA S., BROCK D.L., ASHTON K., *The Networked Physical World : Proposals for Engineering the Next Generation of Computing, Commerce and Automatic-Identification*, October 2000.

[WEI 03] WEIS S.A., SARMA S., RIVEST R., ENGELS D., *Security and Privacy Aspects of Low-Cost Radio Frequency Identification Systems*, 2003.

PART FIVE

Middleware

Chapter 7

Middleware for the Internet of Things: Principles

The concept of the Internet of Things aims at connecting everyday objects with each other. The goal is to enable the emergence of new distributed applications for automatically managing a large numbers of objects like a supply chain management. This automation seeks to improve reactivity and reliability of systems involving a large number of objects, since these objects are becoming more and more difficult to control by human operators. Applications, commonly proposed as an example, cover various fields such as logistics, remote medicine, electronic ticketing, access control, home cares service, etc.

The objects may be related more or less directly to application services or even users. This particular type of interaction aims to allow a system or an individual to retrieve information from an object as well as information about it (in particular its identity in the case of RFID tags) and, in turn, to be able to give instructions within an automated framework.

The middleware implemented above for different equipment plays, in this context, a key role. In general, its infrastructure makes it possible to constitute a network whose terminals (objects materialized by their RFID tag), have low computing power and little or no energy resources. Due to low computing power and inadequate energy resources, the tags face problems similar to the one of embedded and mobile systems, for example sensor networks. Intelligence in case of the Internet of Things is located

Chapter written by David DURAND, Yann IAGOLNITZER, Patrice KRZANIK, Christophe LOGE and Jean-Ferdinand SUSINI.

in the distributed infrastructure and is mainly supported by middlewares, contrary to the original Internet based on intelligent terminals.

The goal of middleware is to allow the development of new information systems by integrating numerous, varied and distributed existing resources while reducing data acquisition time, providing a flexible and scalable infrastructure, making the heterogeneity transparent and allowing the context and user adaptation.

Middlewares seek to hide the information in tags from users (people or software). A first solution is to highlight the service notion while leaving aside the underlying architecture. The middleware must then facilitate discovering, searching and accessing of services. Another solution is to consider all the tags as elements of a distributed database. The middleware must in this case allow access to data and provide means for constructing and evaluating the requests in a distributed manner.

On the one hand, the middleware must be capable of handling of volumes of data, taking into account the possible proliferation of considered objects. On the other hand, as objects, actually connected to the infrastructure at an instant t, say for a quite short duration, the middleware must satisfy relatively strict timing constraints in terms of reaction in an end-to-end chain.

In the following three chapters, we will develop a state of the art of middlewares which could be used for the Internet of Things. There are a wide variety of information systems based on the integration of resources (data and programs). Current technologies that meet these needs are all varied. Therefore, the promoters of RFID solutions have to adapt existing middlewares to the problem of the Internet of Things. As we just noted, two major complementary approaches emerge and contribute to split the middlewares into two main sets: data-oriented and service-oriented. In all cases, the goal of middleware is to provide high-level abstractions to filter and aggregate information. The reader should not be surprised if it is less often directly referred to the RFID tags in the subsequent chapters. In the first chapter, we provide a presentation of principal concepts of middleware as they were conceptualized in the early 1990s. This chapter will allow readers to get used to the basic concepts and understand the issues of the Internet of Things. The reader may refer to [KRA 08] for a general and complete description of middlewares. In Chapter 8 we will present standardization efforts that have been conducted in this area by analyzing in particular the standards which are considered as the most suitable for the Internet of things. Chapter 9 will present the state of the art of industrial solutions, besides more exploratory and/or academic proposals to guide the future developments of the Internet of things.

Finally, we conclude by presenting different future perspectives and an overview of current trends in this area.

7.1. Distributed applications

This section aims at providing the reader with necessary knowledge about middleware to facilitate its comprehension in subsequent parts of the book.

7.1.1. *Principles*

The term *distributed application* designates a set of, at least, two processes that cooperate through a communication system, by exchanging information according to a defined protocol. These processes may be on the same machine, but they can also be distributed in many physical machines.

With the development of communication networks, this type of application has become essential to meet many needs, including data centralization and backup, specific services requirement and execution, load balancing for scientific computing, geographically remote applications connection, cooperative work, etc.

Faced with the multitude of needs, the tendency is to simplify the design, the implementation and the maintenance of such applications. For this, various models of architectures have been built. Compared to their origin, these architectures have largely evolved in such a way to mask, among others, the specificities of host system and the location of distributed items.

7.1.2. *Client-server model*

The simplest model of distributed application is the *Client-Server* model. This model runs a process called *Server* performing on-demand processing at the request of one or more processes called *Client*.

Figure 7.1 illustrates this principle: a *Client* process, hosted on machine *A*, sends a request to a *Server* process, hosted on machine *B*. This one returns the result of the request.

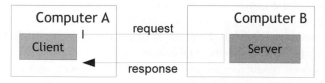

Figure 7.1. *Interaction in the Client-Server model*

The *n-tiers* model is an extension of *Client-Server* model in which a *Server* process may, in turn, become the *Client* of another process during the request processing. With the increasing development of communications, particularly the Internet, this model is being widely used.

Figure 7.2 illustrates how an application is designed in *n-tiers*, based on an example commonly used on the Internet. A *browser* running on machine *A*, sends a request to an *HTTP server* hosted on machine *B*. The protocol used between these two entities is HTTP HTTP[1] [FIE 99]. Then a module of the *HTTP server* communicates with a database server, hosted on machine *C*, this time by using its proprietary protocol. The HTTP service thus plays in both the server role as regards the browser and the client role as regards the database server. With this example, we can talk about *3-tiers* architecture since the chain between processes has length of three components.

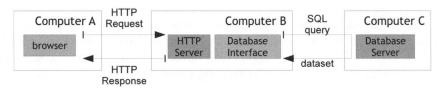

Figure 7.2. *Interactions in a n-tiers model*

In this kind of applications, the transmission and reception of a request based on the mechanism of *Sockets* [BES 87] are widespread. This solution enables the transmission of data between two separate machines in a network through a communication channel, using low-level primitives. It offers the advantage of being supported by almost all operating systems and programming languages, however it restricts designers to a work overload caused by the code related to the transmission, the reception and the error control of processed data.

7.2. RPC: Remote Procedure Call

The various physical communication protocols (TCP/IP, UDP/IP, *ATM*, etc.), the heterogeneity of operating systems (POSIX, *Win32* layers), their versions and libraries, the different hardware architectures (CISC, RISC), data representations (Big Endian, Little Endian) and the variety of programming languages make the implementation of distributed applications complicated and can be a source of errors. In order to facilitate the development, the RPC[2] mechanisms emerged in the 1980s [BIR 84]. Based on Sockets, this technology allows the establishment of a mechanism for remote procedure calls in a transparent manner, by taking care of low-level communication aspects. It also ensures the translation of data into a comprehensible representation by heterogeneous

---

1 HTTP: *Hyper-Text Transfer Protocol.*
2 RPC: *Remote Procedure Call.*

platforms, thanks to XDR[3], a data representation protocol independent of the hardware architecture.

The principle of the RPC, as shown in Figure 7.3, starts with the description of services provided by the server process in a defined language (RPCL), followed by the compilation of this description into source code for client and server programs. These generated files, called stubs, include code for the establishment of service and communications. The designer has to write the processing procedures codes.

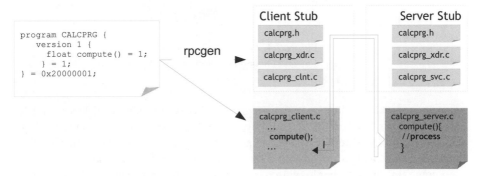

Figure 7.3. *Code generation using RPC*

7.3. Object-oriented middlewares

Like RPC, the middleware represents a set of software layers that can facilitate the interaction between remote processes and mask some details related to the application execution environment like the operating system.

The application creation process is quite similar to the RPC technology, namely, the definition of services provided by an object, which is then followed by a generation of stubs for the establishment of communication process between various entities of the application.

But if RPC is mainly used with procedural programming languages, middleware extends the notion of remote procedure call to the notion of remote object manipulation.

In terms of code, the use of an object-oriented approach allows the reduction of errors, in particular due to strong data typing and to easy-to-handle mechanisms of error management and propagation.

In addition, the middleware generally provides supplementary properties and concepts which the designer can use to build the application and facilitate the deployment, some of which we can mention:

---

3 *XDR: eXternal Data Representation.*

– *the separation of interfaces and implementation of the object* and the interface concept allows the designer to have access to services irrespective to their implementation. This principle allows the update and the extension of service implementation functionalities having the client having to change the way to access the service (Figure 7.4);

Figure 7.4. *Interface separation and the implementation*

– *the transparency in the communication protocol:* the low level layer used for communications is taken into account by the middleware. Stubs generated from the service definition ensure the physical linkage between the implementation that provides the service and the object that uses it. Thus, the application design is independent of the used communication technology used. It does not suffer from any change in support even if the middleware supports it;

– *the transparency in the location of objects:* using the mechanisms such as the *naming service*, remote objects can be located without resorting to physical addresses of nodes on which they are hosted.

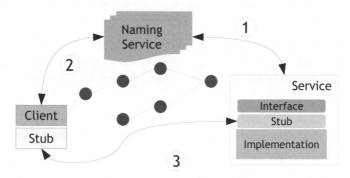

Figure 7.5. *Service research using a naming service*

In the example shown in Figure 7.5, a *service* registers with the *naming service* (1). A client, when he wants to use such a service, will send a request to the naming service (2). It localizes the instance of the service and connects the client and server stubs (3). The interactions between the two objects can then begin.

7.3.1. *Examples*

There are different object-oriented middleware architecture implementations. Among the most common ones, we can cite CORBA[4] of the OMG[5] [OMG04], Java RMI[6] of Sun Microsystem [SUN 04], or *Framework .Net/Remoting* of Microsoft [MIC 09]. The latter tends to gradually replace the architecture DCOM[7] [MIC 96].

In principle, these three architectures operate in a similar manner and we can find most of the concepts discussed above. The differences arise, from our point of view, mainly in the way of specifying service functionalities, to generate intermediate layers or to find remote objects.

7.3.1.1. *CORBA architecture*

CORBA is a middleware specification standardized by the OMG. The ORB[8], the core of this architecture, ensures the automation of communication tasks, object localization and activation, and translation of messages exchanged between heterogeneous systems. Figure 7.6 illustrates this architecture.

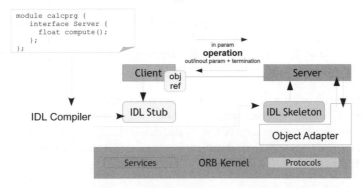

Figure 7.6. *General operating principle of CORBA architecture*

As a specification, CORBA is neither linked to any hardware/software platform or to any particular programming language. Therefore, we can find its different implementations such as OPENORB [BLA 01], MICO [PUD 00], ORBACUS [ION 05] and JACORB [BRO 97].

---

4 CORBA: *Common Object Request Brokker Architecture.*

5 OMG: *Object Management Group.*

6 RMI: *Remote Method Invocation.*

7 DCOM: *Distributed COM.* COM: *Component Object Model.*

8 ORB: *Object Request Broker.*

The CORBA has an associated interface definition language (*IDL*[9]) which makes it possible to describe the services provided by an object, independent of the programming language used for its implementation. An IDL compiler can then translate an IDL interface to a particular programming language. This step will generate portions of code including the *stub* and the *skeleton*, which ensure the linkage between the execution environment, the implementation of the server, and the client.

In the CORBA world, a server object is considered as an IDL interface instance. Its location is masked by a reference mechanism (IOR[10]) which combines to one or more access paths by which a client can contact it.

7.3.1.1.1. Definition/generation phase

The service declaration is described in an interface in OMG-IDL language. There are IDL compilers for various programming languages, including C++, Java, Python or Ada. On the server side, the implementation is usually built by means of the *heritage of generated Skeleton*. On the client side, a *Stub* is used to transmit requests to the server.

7.3.1.1.2. Establishment of a service and invocation

After its instantiation, a server object must be registered and activated within the ORB to be ready to receive client requests. It can choose to use the *Naming Service* to be referenced by a name describing it. The customer will use this same service to retrieve a remote reference of the server object.

During the invocation of a service on the server object, the request sent by the client goes through, in a transparent maner to the *Stub*, which translates possible parameters in a portable data representation (CDR[11]) and sends them to the ORB core. The ORB Core will transmit requests from the client to server and return the result of this request on the reverse route.

The *Object Adapter* is responsible for routing requests coming from the ORB core. It searches the Skeleton corresponding to the destination of the request, which translates parameters in the CDR format back to the format of the target object and then invokes the specified operation on the server object.

The different parts of the ORB communicate between themselves through the GIOP[12] protocol. This generic protocol also allows, from the version 1.2 of specifications, the communication between ORBS from different vendors. The standard

---

9 IDL: *Interface Definition Language.*
10 IOR: *Inter Operable Reference.*
11 CDR: *Common Data Representation.*
12 GIOP: *General Inter ORB Protocol.*

implementation of GIOP is performed in the form of IIOP[13]. This protocol is used when the entities of the application are distributed over different machines, but also when they are hosted on the same machine.

7.3.1.2. *RMI architecture*

RMI is an RPC-based architecture, dedicated to the Java platform. It is object-oriented and supported by all environments using a JVM[14]. The problems related to data encoding and are inherently masked by the nature of Java itself.

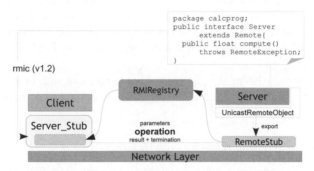

Figure 7.7. *General operating principle of Java RMI*

RMI does not define the IDL language but reuses the proper Java interfaces. Figure 7.7 give an overview of the general architecture of RMI.

An RMI compiler (*rmic*) is used to generate the stubs on the client side or both client and server (Skeleton) sides according to the mechanisms used over the different versions. From version 1.5 of the Java platform, this step can be omitted and the Stub and Skeleton can be generated on the fly during the execution by the JVM.

The *RMIRegistry* service is the naming service of RMI. It allows remote objects to be referenced in a directory and allows clients to find the references of remote objects using their associated service name.

7.3.1.2.1. Definition/generation phase

The declaration of remote services is done by using a Java interface which extends, directly or indirectly, the interface *java.rmi.Remote*. A server object must implement this interface, so as to be used with RMI. Optionally, it can also extend the class *java.rmi.server.UnicastRemoteObject*, which suggests some methods for automating tasks of linking with RMI bus. From this implementation, the (rmic) compiler generates a stub that will be used on the client side.

---

13 IIOP: *Internet Inter ORB Protocol.*
14 JVM: *Java Virtual Machine.*

7.3.1.2.2. Establishment of a service and Invocation

When an instance of server is created, it must be exported under the form of a remote reference. The object *UnicastRemoteObject* provides methods dedicated to this task. During the registration of the server in the *RMIRegistry*, the created remote reference is transmitted together with the name used to refer to the service.

To access the server, the client calls the *RMIRegistry*, to which it transmits the service name under which the server is registered. The remote reference is transmitted to the Stub which is responsible for transmitting the invocations from client to server. The data encoding and decoding is achieved by *serialization* and standard mechanism of the Java platform for the object transmission through any kind of flow.

From the version 1.5 of the platform, there is no more stub generation step since the stubs are created on the fly during the application execution. However this does not change the principle of RMI.

7.3.1.3. *Microsoft.Net and remoting architecture*

Remoting services have been developed to succeed DCOM and facilitate the development of distributed applications in the *Microsoft.Net* platform.

Like the *Java* technology, this architecture is based on its own intermediate code interpreted by a virtual machine. Unlike *Java*, the intermediate code can be constructed from various programming languages including *VB.Net*, *CSharp* and *C++*. Although it is primarily designed for the *Microsoft Windows*, operating system .Net exists now has been implemented in open source projects more or less accomplished (*Mono* [NOV 09]). It can be used on operating systems such as *Linux*, *Solaris*, *BSD* and by extension *MacOS X*. Its portability makes it an interesting multi-platform and multi-language development environment in the context of distributed applications even if we have to await the completion of porting projects to fully exploit *.Net* in a completely heterogeneous environment.

Like RMI, *.Net* does not define additional IDL but can use the interfaces provided by supported object languages to describe the services of a remote object. Figure 7.8 illustrates the operation using *CSharp* as definition language.

The stub generation mechanism on the client and server sides is not necessary; it can be replaced by *dynamic proxies* generated on the fly during the execution.

Currently, there is no naming service to link clients and remote objects. Clients must know the parameters and servers location with which they want to interact. The Server location details have to be directly specified within the code, but they can also be defined in a configuration file regrouping parameters dedicated to communications.

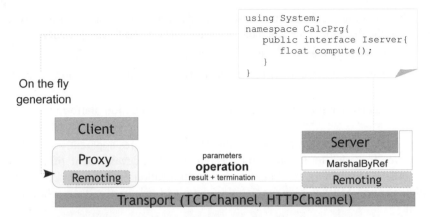

```
using System;
namespace CalcPrg{
    public interface Iserver{
        float compute();
    }
}
```

On the fly
generation

Client

Proxy   parameters   Server
                         operation   MarshalByRef
Remoting   result + termination   Remoting

Transport (TCPChannel, HTTPChannel)

Figure 7.8. *General operating principle of remoting in the*
Microsoft.Net architecture

7.3.1.3.1. Definition phase (SDK 1.1).

The declaration of remote services is done by writing a simple interface in *CSharp*, VB.*Net*, *C++*, or any language supported by the platform. A server object must implement this interface and extend the class *MarshalByRefObject* in the same language as the one of the interface or in another one supported by *.Net*.

7.3.1.3.2. Establishment of a service and invocation

The communication between client and server is done through a communication channel (*Channel*). By default, two types of channels are proposed, the binary-type channel (*TcpChannel*) and the text-type channel, based on the protocol SOAP[15] (*HttpChannel*).

On the server side, the selected channel type and the listening port must be configured with the *ChannelServices*, to prepare the communications channel. Finally, the remote object should be activated by using the services offered by the object *Remoting Configuration* and specifying among other things, the name to which it must be referred.

On the client side, the same communication channel must be configured before getting a remote reference to the server object, by specifying its connection address.

7.4. Summary of object-oriented middleware architectures

We have presented an overview of three object-oriented middleware architectures. We have seen that they are based on the same basic principles to facilitate the development of distributed applications:

_____

15 SOAP: *Simple Object Access Protocol*.

– by providing an abstraction of the services offered by an object, primarily due to thanks to the decoupling between the service interface and its implementation;

– by automating the establishment communication among the distributed objects;

– by ensuring, in a relatively flexible manner, the transparent location of distributed services;

– by standardizing the technical layers related to execution platforms.

In short, whatever they are, the middleware architectures apparently have the same objective. In order to facilitate the design and the development of applications (potentially distributed), they aim at offering application designers and programmers with high level API[16] to hide the technical details of low-levelAPIs on which they are based.

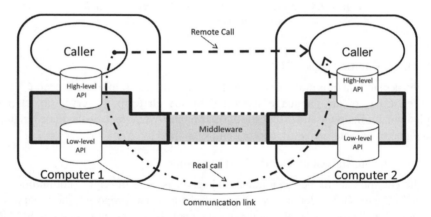

Figure 7.9. *Middleware*

As shown in Figure 7.9, the middleware is a software layer *interposed* between the application objects (here the *invoking object* and the *invoked object*). It establishes the concepts and mechanisms allowing the two objects to be linked and interact together using the low-level APIs provided by support systems. At the same time, it offers high-level APIs to application objects. With the high-level APIs, the application designer/programmer can devote his energy and focus on his application logic rather than having to get lost in the maze of technical details.

Following the ISO standardization, Figure 7.10 summarizes the general structure in a large number of middlewares. It is not necessary in the context of this book to present all of its constituents. It is important to understand the role of the three layers of one of its constituents called ODP *channel*. It is structured in three logical layers:

---

16 API: *Application Programming Interface.*

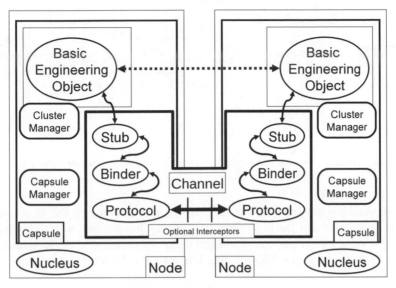

Figure 7.10. *ODP channel*

– the top layer, *stubs layer*, provides the components that allow application objects to be based on the underlying middleware. This layer also provides features related to data formatting and the invocation of services;

– the *link layer* allows application objects to be connected for their interaction. This connection is mandatory before the execution of any interaction;

– the *protocol layer* uses and implements the low level communication mechanisms which allows two parties to exchange information irrespective of the nature of these parties, executing in different processes, are located on the same machine or on different machines.

Like the general structure of middleware explained above which can be found in the main implementations on the market, the notion of *interface* is widely used.

This notion makes it possible to disassociate the type of an object (*X* in Figure 7.11) and the manner in which the services offered by this type are implemented (*X-Impl* in the figure). The interface X contains a set of signatures of operations offered while the implementation of the interface contains the business logic layer of these operations. The notion of interface forms a *contract* between the service provider and service user. In the one hand, the contract stipulates all the services that will have to be implemented by the objects who will exhibit this interface. On the other hand, it specifies all the services to which the user will have access. This decorrelation authorizes the replacement of an interface implementation by another implementation implying minimal changes from the users.

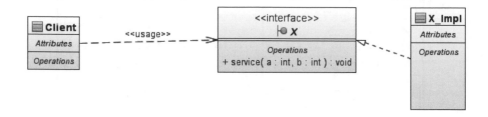

Figure 7.11. *Interface: the contract passed between the service provider and the service user*

Many middlewares provide the property of *transparency to the location*. This property gives the user the possibility to write his interactions with the objects he uses in the same way, regardless of whether these objects are used in the same address space or not. In order to implement this property, the design diagram[17] of *Proxy* is used.

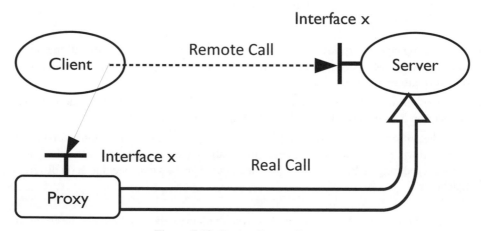

Figure 7.12. *Design Pattern Proxy*

Figure 7.12 illustrates the design diagram which consists of services interface X with which the user wants to interact (*interaction logic*) and a *Proxy* exposing the same interface for user interaction. As *Proxy* and *Server* expose the same interface X; when an invocation is emitted by the user, it will be captured by the *Proxy* (*physical interaction*) which will ensure that it is propagated by the middleware to the concrete implementation of the Server.

---

17 *Design Pattern.*

7.5. The XML revolution

With the success of the Web in the mid-1990s, new types of distributed applications have appeared and created a new enthusiasm. In this context, a new tool for data representation was introduced: the XML language. Developed by the consortium W3C[18], this language did not take long to become the data representation language of choice of distributed applications and led to the emergence of new types of middleware, SOAP, XMLRPC and more generally *Web* services which will take a prominent place in the development of distributed applications.

Today, the access to information systems relies heavily on *Web* technologies. Standardization efforts have allowed the emergence of *Web* services as application development support is available through the *Web*.

In this section we will first provide a brief overview of XML, before presenting the impact that *Web* technologies and XML in particular have had on the structuring of distributed applications, and hence on the nature of middlewares through the concept of *Web* services.

7.5.1. *Overview of XML*

The XML[19] [XML 08] language is a markup language which, like the HTML, is standardized under the W3C consortium. A tag contains a descriptive text whose first component is the tag identifier and is, placed between the characters < and >. The data in a document is framed by tags (an open tag and an end tag). The end tag reuses the identifier in its description the identifier of the open tag preceded by a character /.

However, the language XML differs from the HTML language (which has a limited set of tags describing main aspects of visual representation of documents) in the way that XML is a meta-language which makes it possible to define new tags to isolate all the elementary information that we want to structure, communicate or display. The source code of Figure 7.13 shows an example of a document that handles a list of books using the MODS format [MOD 09] (which is an XML dialect).

This example shows an XML document that begins with a particular tag called prologue, describing mainly the version of the XML specification (version 1.0) and the text encoding (UTF-8). Then we have a tag opening the first element which is the root element of the document. The document has a tree structure. This tree consists of elements nested into each other (with a parent-child relation) and adjacent elements.

---

18 *World Wide Web Consortium*: controls the development and standardization of technologies used on the World Wide Web.

19 EXTENSIBLE MARKUP LANGAUGE.

The prologue may possibly contain the document type definition (DTD). It is also possible to add processing instructions of which the most common is the stylesheet declaration. This stylesheet is used for formatting the display. *modsCollection* is the Document Element in the absence of a DTD. An attribute XMLNS can be added to specify an XML schema as explained in the following next subsection. The example consists of two MODS elements. In each one, we find these elements and common sub-elements (`titleInfo`, `title`, `typeOfResource`, and `identifier`). Some open tags at the beginning of an element have attributes that complete the specification of the element. For example, the `identifier` element has an attribute `type` whose value is the string of characters `"citekey"`.

The next section presents the various efforts of typing in XML in order to validate data presented in an XML document

7.5.2. *Definition of the structure of an XML document*

Sometimes it may be necessary to specify the tags and attributes to which we have rights when writing an XML document, for example if you want to share the same type of document with a community of other writers. Two solutions are possible: DTD[20] or XML schemas.

7.5.2.1. *DTDs*

DTDs are based on the SGML standard [GOL 90] and present some disadvantages:

– DTDs are not in the XML format. Therefore it is necessary to have a specific tool for handling this type of document, which is different from the one used for editing the XML document;

– DTDs do not support namespaces. In practice this means that it is not possible to import definitions of tags defined in an XML document defined by a DTD;

– data typing is extremely limited.

7.5.2.2. *XML Schemas*

XML schemas have been designed to overcome the shortcomings of DTDs and to bring new features beyond those provided by DTDs:

– a new mechanism for data typing is introduced, which allows the management of Boolean, integer, time intervals, etc. It is also possible to create new types from existing types;

– there is a notion of legacy. An element can inherit the content and attributes of another element;

---

20 *DTD: Document Type Definition.*

```
<?xml version="1.0" encoding="UTF-8"?>
<!-- The above line constitutes a prologue -->
<!-- This tag defines a comment -->
<!-- The following tag defines the document root -->
<modsCollection xmlns="http://www.loc.gov/mods/v3">
 <mods>
  <titleInfo>
   <title>Hypertext Markup Language - 2.0</title>
  </titleInfo>
  <name type="personal">
   <namePart type="given">T.</namePart>
   <namePart type="family">Berners-Lee</namePart>
   <role>
    <roleTerm authority="marcrelator" type="text">
      Author
    </roleTerm>
   </role>
  </name>
  <name type="personal">
   <namePart type="given">D.</namePart>
   <namePart type="family">Connolly</namePart>
   <role>
    <roleTerm authority="marcrelator" type="text">
      Author
    </roleTerm>
   </role>
  </name>
  <typeOfResource>text</typeOfResource>
  <identifier type="citekey">HTML1866</identifier>
 </mods>
 <mods>
  <titleInfo>
   <title>
    Extensible Markup Language (XML) 1.0
   </title>
  </titleInfo>
  <typeOfResource>text</typeOfResource>
  <identifier type="citekey">XML08</identifier>
 </mods>
</modsCollection>
```

Figure 7.13. *Example of an XML document describing a bibliography*

– a schema supports a differentiated namespace notion. This allows any XML document to use the tags defined in any given schema;

– indicators of occurrences of elements can be non-negative numbers (in a DTD, it is limited to 0, 1 or an infinite number of occurrences for an element);

– schemas can be designed in a modular fashion.

An XML schema is an XML document. We can handle a schema in the same manner as any other XML file, and in particular to adjoin to it a stylesheet. It is also possible to automate the creation of a document from a schema based on comments and explanations therein.

We can also divide a schema into several parts; each part is responsible for defining precisely a particular sub-domain of the schema. It is possible to include multiple schemas in one scheme, using the element xsd:include. This allows us to modularize a schema and thus easily import some parts from existing schemas or to create another part outside the basic schema.

The following sections are devoted to the description and implementation of *Web* services.

7.5.3. *Web services*

With the growth of XML, we observe the emergence of a new type of distributed applications built on *Web* technologies, leading to the notion of *Web* services. The W3C defined itself as *Web* services and is a distributed software system designed to support interactions between different machines which are interoperable due to the widespread adoption of the Internet standards on most architectures. A *Web* service has an interface described in a standardized comprehensible format by the machine (mainly the WSDL[21] explained in the section 7.5.4). Other systems interact with the *Web* services using SOAP[22] messages (described in section 7.5.6) or XML-RPC[23] [WIN 99], themselves usually built over HTTP. The use of HTTP (but other Internet protocols can be used) has the advantage of being broadly adopted and generally well taken into account by the firewalls of organizations. The data encoding is done in an XML document.

One of the important points during the development of Web services is that programming languages are totally independent of standards. The support for different programming languages is not specified and each provider of technology solutions is thus free to propose different programming models if the standards are respected. Thus

---

21 *Web Service Description Language*.
22 *Simple Object Access Protocol*.
23 XML Remote Procedure Call.

the generation of messages from a particular programming language is absolutely not normalized, contrary to, for example, what happens in the case of CORBA. As a result, applications are designed as a set of business services. Each service is developed for a specific task which is well defined and business oriented, regardless of the applications that use it.

An alternative approach has achieved significant success in *Web* services domain: the approach REST[24] [FIE 00]. This approach describes a particular of architecture style for distributed applications which is applied in particular to the case of *Web* services. A number of standards appear in this field but we will not describe them here. In practice the so-called *RESTful* web services are based directly on the use of the HTTP protocol and associate a different URL with each operation. The messages are encoded in XML or JSON[25] [CRO 06]. This results in a much simpler implementation, which does not require special infrastructure, hence many online services prefer this approach.

With the simple feature of implemented technologies and the standardization of description and communication mechanisms, *Web* services can be used in the context of inter-company exchanges. Many online services (paid or unpaid) can be used within different applications. A good illustration is the *GoogleMap* service [GMA 09] which is very often reused in many applications using geo-locations such as WIFI access points.

In the following sections we will introduce some key technologies of the Web services domain, starting with *Web* service description, then the tools for the location on the network (directory), and communication mechanisms in particular the SOAP protocol.

7.5.4. *Description of Web services-WSDL*

Web services are accessible through an interface that is defined by a standard mechanism. In many cases this description is based on WSDL [CHI 07] which is a dialect of XML. This description focuses primarily on functional aspects of a service but other languages can go further and specify behavioral aspects (such as BPEL [JOR 06], WS-CDL [KAV 04], WS-*specifications, etc.).

The WSDL language is built on XML fomat and is used to describe the supported data types and functions offered by *Web* services. WSDL descriptions are independent of the platform and the programming language. It plays an equivalent role with the IDL language of CORBA (section 7.3.1.1).

---

24 *Representational State Transfer.*
25 *JavaScript Object Notation.*

In WSDL, a *Web* service is described by a set of messages that we can exchange with it through access points: ports. The full definition of a port based on:

– a type (*Port Type*): an abstract description of the communication interface, is represented by a collection of operations supported by the port (and especially their signature) and also includes the definition of the messages consumed and generated by the service;

– an association (*Binding*): provides an indication of the protocol used to exchange messages with the service (for example SOAP over HTTP), and the associations between the interface message and the message types supported by the underlying communication protocol;

– a network address to localize the access point.

The result is the structure presented in the document of Figure 7.14.

```
<?xml version="1.0"?>
<definitions>
<types>
    .....
</types>
..
<message>
    .....
</message>
..
<portType>
    .....
</portType>
..
<binding>
    .....
</binding>
..
<service>
    .....
</service>
</definitions>
```

Figure 7.14. *Structure of a WSDL document*

The definition of elements portType corresponds generally to the definition of the *Web* service itself. It describes the operations that can be executed and the exchanged messages. We can find in this definition the notions of interface (*portType*) and method (operations) already presented in the IDL of CORBA. WSDL defines 4 main types of operations:

– *Request-Response*: the operation receives a request and generates a response;

– *One-way*: the transaction receives a message but does not answer;

– *Solicit-Response*: the operation sends a request and waits for a response;

– *Notification*: the operation sends a notification without waiting for an answer.

In practice, WSDL defines associations (*Bindings*) only for Request-Response and One-Way operations. The document in Figure 7.15 illustrates the definition of these two operation types.

```
<portType  name="portTypeName">
      <operation  name="operation1Way">
          <input name="requete"  message="
              tns:MessageInput"/>
      </operation>
      <operation  name="operationReqResp">
          <input name="requete"  message="
              tns:MessageInput"/>
          <output  name="reponse"  message="
              tns:MessageOutput"/>
          <fault  name="exception"  message="
              tns:MessageException"/>
      </operation>
</portType >
```

Figure 7.15. *Port definition example*

Once the port type is defined, we declare an association over a communication protocol as described in Figure 7.16.

```
<binding  name="lienSOAP"  type="portTypeName">
    <soap:binding  style="rpc"  transport="http://
        schemas.xmlsoap.org/soap/http"  />
    <operation  name="operationReqResp">
      <soap:operation  soapAction="http://monservice.
          com/operationReqResp"/>
      <input><soap:body  use="literal"/></input>
      <output><soap:body  use="literal"/></output>
      <fault><soap:fault  use="literal"/></fault>
    </operation>
</binding>
```

Figure 7.16. *Example of an association definition*

7.5.5. *Location of Web services*

In addition to the description of *Web* services, it is also important to be able to, like what is commonly done in systems such as CORBA, discover and locate available *Web* services on the network. For this, in 2001, the UDDI standard[26] [UDD 02] was created under the aegis of OASIS[27] which defines a directory of business-oriented *Web* services, independent of the platforms accommodating them. Different types of information can be stored in directories from simple identification and location information to the publication of technical data or even the WSDL description of the service itself. However, UDDI is not necessarily linked to the use of WSDL, as a unique description standard of *Web* services. The standard specifies a particular format of information (*tModel*) and, provides a set of specifications to exploit the information in the UDDI directory which makes it possible to: (*i*) publish descriptions of registered services (their WSDL description) in dedicated folders (UBI: *Universal Business Registry*); (*ii*) submit requests based on keywords of directories and (*iii*) browse through the descriptions obtained by these requests (Figure 7.17).

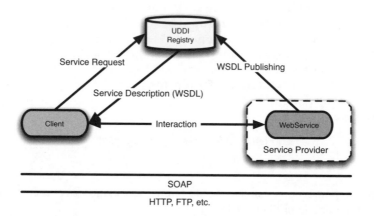

Figure 7.17. *Architecture of Web services*

The directory service has a standardized description in the WSDL format and the service can be invoked through the SOAP protocol.

However, this solution, initially supported by the main actors of the market (IBM, BEA, *Microsoft*, etc.) is directly in competition with search engines on the Internet and existing directories in organizations (Active Directory, LDAP, etc.). Accordingly, relatively few systems effectively support this solution.

---

26 *Universal Description Discovery and Integration.*
27 *Organization for the Advancement of Structured Information Standards.*

7.5.6. *SOAP*

SOAP is a communication protocol commonly used in *Web* services. It has resulted from the standardization efforts of a remote procedure call mechanisms, in particular of XML-RPC [WIN 99] and [RHO 09]. It ignores the application-level protocols over which it is implemented and executes over both HTTP (the most common) and other protocols such as SMTP. Thus, SOAP provides the communication service over which *Web* services can be built. SOAP defines:

– *SOAP Encoding Rules*: the encoding of messages and data with a set of encoding rules for data types defined by applications;

– *SOAP RPC Representation*: the method used to perform the exchange over other application protocols;

– *SOAP envelope*: the content of messages, and the encoding of operation names, the entities dealing with different messages, and the optionality of messages (see Figure 7.18).

SOAP defines a standard structure of messages in which the content of the messages is separated from the meta-data related to the exchange of messages. A SOAP message consists of two parts:

– the header indicating the message subject, the description of the sender and the information required to route the message to the receiver;

– the body which can contain:

  - any document,
  - the call of an operation offered by the receiver service with the values for input parameters,
  - the values generated as a result of a call,
  - an error message.

SOAP has been critized that it adds few features as compared to what has already been possible with HTTP and XML. In addition, by combining multiple operations to the same URL (multiplexing), it becomes difficult to use the cache management infrastructure associated with HTTP, which is its important feature. The use of XML is also often criticized because the representation of the exchanged messages is more complex. However, it is possible to encapsulate the objects which are directly encoded in binary to improve the performance. Finally, the abstraction of the underlying communication protocol, in particular the HTTP, makes it impossible to know which method of the HTTP protocol is concretely used to transport data, which can create problems of efficiency/safety.

As the added value of SOAP is low as compared to implementation using directly HTTP, many online services prefer the REST approach.

```
POST /MonService HTTP/1.1
Host: www.monservice.com
Content−Type: text/xml; charset="utf−8"
Content−Length: nnnn
SOAPAction: "MonAction−URI"

<SOAP−ENV:Envelope
   xmlns:SOAP−ENV="http://schemas.xmlsoap.org/soap/
      envelope/"
   SOAP−ENV:encodingStyle="http://schemas.xmlsoap.org/
      soap/encoding/">
   <SOAP−ENV:Header>
     <MessageID
        xmlns="http://schemas.xmlsoap.org/ws/2003/03/
           adressing">
           proto://DefinitiondunmessageID
     </MessageID>
     <ReplyTo xmlns="http://schemas.xmlsoap.org/ws
        /2003/03/adressing">
        <Address>
           http://adressepourlareponse.com
        </Address>
     </replyTo>
   </SOAP−ENV:Header>
   <SOAP−ENV:Body>
        <m:MonAction xmlns:m="MonAction−URI">
           <symbol>DIS</symbol>
        </m:MonAction>
   </SOAP−ENV:Body>
</SOAP−ENV:Envelope>
```

Figure 7.18. *Example of a SOAP message on an HTTP request*

Web services have emerged in recent years as the solution of choice for distributed application deployment on the Internet. Numerous vendors offer complete solutions and their activities remain very strong in this field. Thus will see that these solutions are also very popular in the world of the Internet of Things.

7.6. Middleware for the Internet of Things

Applications of the Internet of Things are inherently distributed: tags and readers are geographically distributed in many different places, potentially too far away from each other as well as from enterprise information systems.

Most of the concepts that we have presented in this chapter consider middleware communication vectors among distributed application entities. But nowadays, this approach is often completed by elements serving to organize enterprise information systems. In this framework, the data structuring, the management of information flows and their processing are as important as communication mechanisms.

It is thus necessary to have a complete intermediate facilitating architecture, on one hand, the communication between RFID tags and on the other hand, the communication between users (human or software). Beside the aspects related to communication, it is also necessary to structure data, exchanges and processing in order to integrate them into enterprise information systems in the best way and enable them to optimize their management.

In this context, as we mentioned earlier in this chapter, there are two distinguished approaches which have become complementary to the proposed solutions to support communications between components of a distributed application.

The first approach is called service-oriented which considers that the objects (things and especially their RFID tags) will provide information to components which provide services to objects themselves, other objects, human operators or other applications. These services can be registered in directories, be searched or discovered by objects or applications. These solutions focus on the structuring of the information system for function of data analysis and processing which are performed according to the business logic of the user. We will present various proposals in this area. This approach (called SOA: *Service Oriented Architecture*) is currently very popular in the design of enterprise information systems and we will see in Chapter 9 that many proposed solutions are based on these principles.

A second approach its called data-oriented proposes, considers the objects and in particular their RFID tags in two different ways: (i) the objects are seen as distributed sensors that contain information consisting of a distributed database in which information changes in time, (ii) the objects are seen as memory cells constituting a distributed tuple space of several kilobytes of hundreds or thousands of passive tags or more advanced devices. We present the elements which, in the solutions for different applications of the Internet of things, correspond to this approach.

7.6.1. *Service-oriented middlewares*

A service-oriented middleware focuses on all issues related to the integration of components capable of performing a particular function within an application (see [MIC 08]). In this point of view, this approach is very similar to the solutions for distributed objects we have described above. The notion of service reuses the principle of exposing the offered services through interfaces, but it differs from the object approach by different aspects, including the notion of contract and the composition, which are both borrowed from component systems. Finally, the notion of service is generally used to describe functional aspects of components attached to business logic in the structuring of an enterprise information system.

These concepts have been popularized by the apparition of *Web* services (see section 7.5.3). However, a service can be either a business component or the establishment of a technical function (as it is often the case in the implementation of platform such as OSGi, see Section 8.3.1).

The service-oriented middleware generally uses an approach for the refinement of high-level business services. Thus, a business function "management of orders" in a service-oriented approach will be refined into sequences of sub services: "creation of an order", "assembling of ordered products", "shipping order", "billing", "cancellation-modification of the order", etc. These sub-services will be chained to describe several different work flows in the function of the sequences of events received by the system.

Due to its great similarity with the distributed object architectures and components, we will focus primarily in this section on the programming model of a system based on service oriented architecture rather than on inter-component communication aspects whose concepts have already been presented in the previous sections. However, it is important to note that the service-oriented solutions do not necessarily infer that applications are distributed, which sometimes cannot be really qualified as middleware in the usual definition.

Structuring the information system is therefore based on the data processing and analysis which are considered as functional services. The middleware aims at optimizing the services flow to extract relevant information on the business conduct of the user company.

A service is a behavior (representing a processing) defined by contract, which can be implemented and provided by a component to be used by other components, on the exclusive base of the contract. A service is accessible via one or more interfaces. An interface describes the interaction between client and service provider. This interaction is defined at a low level by presenting the elementary operations that can be invoked. From an operational point of view, there is the definition of operations and data structures that contribute to the execution of service and to the valid scenarios of sequence of elementary operations. Services are loosely coupled, that is to say, the establishment of a service from an outside view depends only on the contracts that it engages to provide and the contracts it uses. To strengthen this loose coupling, the provision of a service involves two types of interfaces:
– the required interfaces (client side);
– the provided interfaces (provider side).

The contract specifies the compatibility (compliance) between these interfaces. Beyond the interface, every part is a *black box* for the other (encapsulation principle). By consequence, the customer or the provider can be replaced as long as the substitute respects the contract (is compliant). The compliance of interfaces can be performed by mediators that edit data flows. Unlike the technical components, a service is unique and does not have multiple instances. It can be localized and the granularity from the functional point of view is usually wider than the one of a technical component.

In the case of the Internet of Things, the operations supported by this type of middleware are:

– transformation of raw events from RFID readers directly into application events which are normalized and accessible on the network;

– aggregation and filtering of data, and especially application events themselves;

– finally, exchanges with other business applications.

Service-oriented architectures have been focused to respond with flexible and adaptable approaches to the problem of the evolution and restructuring of enterprise information services and in particular inter-organization exchanges. In the context of the Internet of things and particularly its applications to logistics, we find a solution adapted to the integration of RFID tags into information systems and to the interaction between various organizations participated in the global chain. Therefore, a large number of solutions considered in Chapter 9 are based on service-oriented architectures.

7.6.2. *Data-oriented middleware*

The data-oriented middleware proposes to consider RFID tags and the objects with which they are associated as the memory cells used for storage or data production. The middleware will consider in this case RFID tags as elements of a distributed database and will strive to enable access to data and to provide effective interrogation means of the distributed database. Section 7.6.2.1 will present approaches based on the notion of tuple-space. Section 7.6.2.2 will present approaches based on distributed databases by adopting the point of view of embedded systems (sensor networks, smart cards, etc.).

7.6.2.1. *Tuple-space-based middleware*

This is an implementation of associative memory paradigm for parallel or distributed computing. It provides a repository of tuples that can be concurrently accessed. Thus, if one group of processors produces pieces of data and another uses this information, the producers post their data as tuples in the shared space, and consumers find them following a certain scheme. This is also known as the blackboard metaphor.

This paradigm was highlighted by the mid-1980s at the University of Yale. This is the base of coordination language LINDA [CAR 89] which proposes a general model of parallel computation based on distributed data structures. The tuple-space is a shared data space.

LINDA involves running distributed programs, which do not know each other, and using a shared memory, as the problem is the synchronization. The operations include reading, removing and writing in the tuple-space.

Today, an implementation used tuple-space is *JavaSpaces* [FRE 99], which is included in the JINI technology [JIN 09] (discovery, research services, event and transaction concepts). In fact, *JavaSpaces* is the Java object version (inheritance, methods, etc.) of LINDA tuples. *JavaSpaces* may produce for example notifications reacting to a simple input (members equality, inheritance, etc.) which is formalized by the notion of contract. The *JavaSpaces* proposes mechanisms for distributed persistence and data exchange that simplifies the design of distributed systems by enabling the implementation of distributed algorithms based on the movement of objects between participants and not on the remote invocation. A *JavaSpaces* contains entries that correspond to Java classes.

From the point of view of the Internet of things, this notion of tuple-space may be interesting, because in a data-oriented middleware it is interesting to have a shared memory space used for storage of all data accessible through RFID readers. This space may be unique, so that each node can connect to it, which results in a significant load on the space, but an easier coherence management. The space can also be duplicated on each node, thus distributing the load more. However, the coherence is much more difficult to maintain. Having a shared space corresponds to a primary need of RFID applications, which is the intelligent data storage. This model allows the abstraction of information collecting operations and the definition of filtering and aggregating information operations through the access to shared space.

We will see the implementation of this model in some solutions described in Chapter 9, especially those based on JINI technology.

7.6.2.2. *Embedded databases*

Another data-oriented approach to organize distributed applications is to consider systems as a distributed database. By the way of tuple-space-based model presented in the previous section, the idea is to have a distributed data space that can be interrogated using requests as defined for classical information systems. From the definition of global requests, which represent the processing (analysis, aggregation, matching, correlation, etc.) that an application want to perform on data, the middleware will automatically derive a set of distributed local (local processing on acquired or read data) processes, their methods of communication (communication network, exchange structuring, etc.), and the flow of information sent to the network in order to satisfy the global demand. A number of constraints will be integrated to optimize network operations (memory consumption of cells, energy consumption, performance, reliability, etc.).

This approach has achieved significant success in the deployment of applications based on sensor networks. Currently, some work extends this approach to RFID tags. In the case of active tags, which can, for example, embed a temperature sensor, we deal with problems very similar to sensor networks. In the case of passive or active tags with a small amount of information storage, we can consider each tag as an information cell of the database. From now, the middleware will have to focus on the distribution of information capturing and writing process since the tags are not always available (present in front of a reader). In section 9.5 we present some works in this field.

This type of middleware that integrates both aspects of information communication and data structuring has achieved a major success in the field of embedded systems. These concepts have been proven in the field of sensor networks and smart cards. The fact of considering the specific problems of embedded systems makes these technologies particularly adapted to the problems of RFIDtags. This tends to open the field of perspectives of the RFID application in new types of applications.

7.7. Conclusion

Middleware plays a key role in the development of distributed applications. Applications of the Internet of things have similar problems. The success of many proposed applications taking advantage of the RFID-based technology is based on cooperation and interoperability between different information systems from different organizations. This is particularly the case for logistic applications such as those highlighted by the EPCglobal consortium. In order to facilitate the cooperation and interoperability, we will see in the next chapter the standardization efforts that have been conducted recently.

7.8. Bibliography

[BES 87] BESAW L., Berkeley UNIX System Calls and Interprocess Communication, 1987.

[BIR 84] BIRRELL A., NELSON B., "Implementing Remote Procedure Calls", *ACM Transactions on Computer Systems*, vol. 2, no. 1, p. 39-59, January. 1984.

[BLA 01] BLAIR G., COULSON G., ANDERSEN A., BLAIR L., CLARKE M., COSTA F., DURAN-LIMON H., FITZPATRICK T., JOHNSTON L., MOREIRA R., PARLAVANTZAS N., SAIKOSKI K., "The Design and Implementation of OpenORB v2", *IEEE DS Online, Special Issue on Reflective Middleware*, vol. 2, no. 6, 2001.

[BRO 97] BROSE G., "JacORB: Implementation and Design of a Java ORB", *Proceedings of DAIS'97, IFIP WG 6.1 International Working Conference on Distributed Aplications and Interoperable Systems*, Cottbus, Germany, 1997.

[CAR 89] CARRIERO N., GELERNTER D., "Linda in context", *Commun. ACM*, vol. 32, no. 4, p. 444-458, ACM, 1989.

[CHI 07] CHINNICI R., MOREAU J.-J., RYMAN A., WEERAWARANA S., Web Services Description Language (WSDL) Version 2.0 Part 1: Core Language, W3C Recommendation, June 2007, http://www.w3.org/TR/wsdl20/.

[CRO 06] CROCKFORD D., The application/json Media Type for JavaScript Object Notation (JSON), IETF - RFC, June 2006, http://www.ietf.org/rfc/rfc4627.txt.

[FIE 99] FIELDING R., IRVINE U., GETTYS J., MOGUL J., FRYSTYK H., MASINTER L., LEACH P., BERNERS-LEE T., Hypertext Transfer Protocol – HTTP/1.1, IETF - RFC, 1999, http://www.ietf.org/rfc/rfc2616.txt.

[FIE 00] FIELDING R., Architectural Styles and the Design of Network-based Software Architectures, PhD thesis, University of Califormia, Irvine, USA, 2000.

[FRE 99] FREEMAN E., HUPFER S., ARNOLD K., *JavaSpaces Principles, Patterns, and Practice*, Addison-Wesley Professional, June 1999.

[GMA 09] Google Maps API Concepts, Google Code Documentation, 2009, http://code.google.com/apis/maps/documentation/.

[GOL 90] GOLDFARD C.-F., *The SGML Handbook*, Oxford Clarendon Press, 1990.

[ION 05] IONA Technologies PLC, "Orbacus Technical Review", 2005.

[JIN 09] Jini Specifications and API, Sun Microsystems - Product, 2009, http://java.sun.com/products/jini/.

[JOR 06] JORDAN D., EVDEMON J., ALVES A., ARKIN A., ASKARY S., BLOCH B., CURBERA F., FORD M., GOLAND Y., GUÍZAR A., KARTHA N., LIU C. K., KHALAF R., KÖNIG D., MARIN M., MEHTA V., THATTE S., VAN DER RIJN D., YENDLURI P., YIU A., Web Services Business Process Execution Language Version 2.0, Public Review Draft, 23th August, 2006, 2006, http://docs.oasis-open.org/wsbpel/v2.0/.

[KAV 04] KAVANTZAS N., BURDETT D., RITZINGER G., FLETCHER T., LAFON Y., Web Services Choreography Description Language Version 1.0, W3C Working Draft, December 2004, http://www.w3.org/TR/2004/WD-ws-cdl-10-20041217/.

[KRA 08] KRAKOWIAK S., COUPAYE T., QUÉMA V., SEINTURIER L., STEFANI J.-B., DUMAS M., FAUVET M.-C., DÉCHAMBOUX P., RIVEILL M., BEUGNARD A., EMSELLEM D., DONSEZ D., "Intergiciel et Construction d'Applications Réparties", http://sardes.inrialpes.fr/ecole/livre/pub/main.pdf, juin 2008.

[MIC 08] MICHAEL B., *Introduction to Service-Oriented Modeling. Service-Oriented Modeling: Service Analysis, Design, and Architecture.*, John Wiley & Sons, 2008.

[MIC 96] Microsoft Corp., DCOM Technical Overview, MSDN Library Specification, 1996.

[MIC 09] Microsoft Corp., Microsoft.NET Framework, 2009, http://www.microsoft.com/net.

[MOD 09] The Library of Congress, Metadata Object Description Schema, 2009, http://www.loc.gov/standards/mods/.

[NOV 09] Novell Inc., Mono 2.4, 2009, http://www.mono-project.com.

[OMG04] Common Object Request Broker Architecture (CORBA/IIOP) – version 3.0, *Object Management Group*, formal document, 01, 2004.

[PUD 00] PUDER A., RÖMER K., *MICO: An Open Source CORBA Implementation*, KAUFMANN M. (ed.), 2000.

[RHO 09] RHODES K., "XML-RPC vs. SOAP", Webpage, September 2009, http://weblog. masukomi.org/writings/xml-rpc_vs_soap.htm.

[SUN 04] Sun Microsystems - JSR Java Remote Method Invocation Specification – rev. 1.1.10, 2004.

[UDD 02] UDDI Standard set, OASIS Consortium, 2002, http://www.oasis-open.org/specs/ index.php#uddiv2.

[WIN 99] WINER D., XML-RPC Specification, UserLand Software, Inc, 1999, http://www. xmlrpc.com/spec.

[XML 08] W3C Consortium, Extensible Markup Language (XML) 1.0, 5th edition, 2008, http://www.w3.org/TR/REC-xml.

Chapter 8

Middleware for the Internet of Things: Standards

Following the work of *Auto-ID Lab*, the development of the EPCglobal consortium provided an opportunity to understand the concept of Internet of Things during a maturation process, leading to a number of standards. The establishment of standards respected by the greatest number of participants guarantees interoperability and persistence of solutions. Therefore in this chapter we will present some of the standards which are currently the most commonly used in the domain of the Internet of Things.

From middleware's point of view, the previous chapter provided an opportunity to present a number of concepts and standardization efforts (CORBA, XML, *Web* services, etc.). This chapter is intended to complement the description of different standards in force with specifications commonly used in the solutions which we will further study in Chapter 9.

For several years the field of enterprise information systems has been the subject of numerous standardizations. The reasons are many. Enterprise information systems are often interconnected. It allows us to take into account *Business to Business* (B2B) problems which require information transfers between a client and a provider. In addition, the integration of different business applications in an efficient business information system requires a standardization of design methods and inter-application exchanges to facilitate the development and maintenance of these increasingly complex systems. In this context, we will present in this chapter only those standards that exist in various proposed solutions to design and integrate applications based on RFID tags.

Chapter written by Yann IAGOLNITZER, Patrice KRZANIK and Jean-Ferdinand SUSINI.

Initially, we briefly recaptulate the standards used for the design of middleware for Internet of Things as proposed by the EPCglobal consortium. Secondly, we discuss the proposed standards for message-oriented middleware based on JMS standard. Finally, we look at standards for service-oriented middleware.

8.1. EPCglobal application environment

The EPCglobal standards have been studied in detail in Chapter 6 and the objective of this section is only to briefly recall the elements of EPCglobal standards which directly impact the design of middleware built on these solutions.

From the middleware perspective, EPCglobal standards propose a number of specifications which are presented below.

The interface for data collection and filtering: ALE transforms raw data from RFID readers in application events. This transformation takes place "in real time". The processing operations are applied to a very large set of cases in which tags are inventoried. ALE provides the means to one or more client applications to acquire EPC data from one or more data sources which isolate client applications and the underlying hardware. The data are assembled into logical units: events, by ignoring the fact that there are one or more readers, and one or more antennas deployed to produce this information. This solution also overlooks how devices are deployed to form a single logical data source. The ALE specification provides declarative means in client applications to specify the type of processing to perform on EPC data (filtering, aggregation, grouping, counting, differential analysis, etc.). Two modes of event delivery can be specified: punctual or recurrent delivery. The ALE specification authorizes the sharing of data between different applications. From this point of view, it is rather a data-centric middleware approach.

Capture event application: EPCIS accepts input application events from ALE, which can be stored for subsequent processing (history, analysis, etc.). These events are handled by processes managing business logic of applications and can deliver EPCIS data (which are themselves stored by the application and/or sent to other EPCIS client applications). It can coordinate multiple data sources while recognizing individually each EPCIS event. Data sources can include filtered and collected EPC data obtained by ALE, and other data generated by devices such as barcodes, human inputs, or data collected by other software systems. Finally, EPCIS applications can trigger actions in the physical environment, including writing (customization) tags, and taking control of other devices.

PML files, constructed from a XML dialect, can gather all information about an RFID tag in a document. Business applications can then use the data contained in this document for business processing.

The location of resources is also standardized using conversion mechanisms of EPC data carried by RFID tags in URI. They are then submitted to ONS servers that can localize EPCIS applications which manage PML documents associated with tags.

Finally, EPC standards describe authentication procedures to secure exchanges between various entities on EPC network.

These specifications manage both data-oriented concern (ALE) and service-oriented concern (EPCIS). All these standards are intended to facilitate the interoperability between different information systems of all companies that are participating in the EPC network and accept sharing some information associated with objects they use and in particular their RFID tags. This is necessary in applications such as logistics, often presented as an example. The security of data access then becomes very important.

8.2. General introduction to message-oriented middleware

Middleware based on the exchange of messages are intended to structure exchanges between various components of a distributed application by proposing a simple and effective communication model: asynchronous sending of messages. This technique has proved its worth and numerous industrial tools are based on standards in this field (MMS, JMS, etc.). Several offers are available on the market: commercial offers such as *MQ Series* of IBM, (which is now called *Websphere MQ*, currently version 6.0) and *MSMQ* of *Microsoft*, free offers and *Open Source*, such as implementations of the *Java Messaging Service* (JMS) specification of Sun Microsystems, *ActiveMQ* of Geronimo group of Apache, JORAM, or *OpenJMS*, the first MOM compatible JMS *Open Source*.

This section will present these solutions, especially messaging built around two main principal specifications: (i) JMS in Java World; (ii) *Jabber*/XMPP which is the subject of an IETF specification whose range of applications extends beyond Java World.

8.2.1. *General instruction to message-oriented middleware*

These are server softwares that offer their customers a Peer-to-Peer communication service by federating message sending and receiving between applications. Communication via messages is an asynchronous communication mechanism widely used to communicate between servers, especially in the context of integration of data and applications in a more global system (EAI, *Enterprise Service Bus*-ESB, etc.). This mechanism is also useful for the implementation of data warehouses, inter-banking messaging (such as AMQ), and more generally information dissemination systems.

The communication by message is usually associated with an associative naming system in which recipients are identified by a part of the message, which is the base of publish-subscribe model that is very commonly used today.

Asynchronous communication systems based on sending messages appear to be particularly suitable (compared to the classical model of synchronous client-server) to manage interactions between loosely coupled systems. Loose coupling can be spatial (geographical distance between entities) or temporal (temporarily disconnected due to mobility or failure). Asynchronous communication models take into account the independence between communicating entities.

MOM are very widespread because they are the technological base for achieving the following applications:

– data integration and application integration (EAI, B2B, ESB, etc.);

– ubiquitous systems and use of mobility;

– monitoring and control of network devices.

The concepts are:

– communicating entities are decoupled. The transmisson (production) of a message is a non-blocking operation. Transmitter (producer) and receiver (consumer) do not communicate directly with each other, but use an intermediate communication object (mailbox);

– two models of communication are provided:

- point to point model in which a message is consumed by a single recipient,

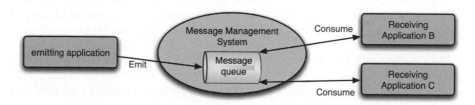

Figure 8.1. *Point to point communication model: producer-consumer.*

- multipoint model in which a message can be sent to a community of recipients (group communication).

In contrast to client-server systems such as CORBA [OMG 09] systems, the development of asynchronous communication systems has been delayed by the lack of standards. These standards remained proprietary for a long time, both in terms of programming model and implementation. In the late 1990s, the emergence of the JMS specification in Java world has partially addressed this handicap. The standard defines a programming model and the corresponding API, but so far no standard has been established for the middleware itself or for the interoperabiliting mechanism as is the case of example of the IIOP protocol in CORBA.

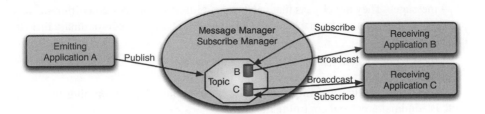

Figure 8.2. *Multi-point communication model: publish-subscribe*

8.2.2. *Java Messaging Service (JMS)*

JMS [?] is a programming interface that specifies a messaging service for sending and receiving messages asynchronously between applications or Java components. JMS implements an architecture of type MOM. Receiving messages in a synchronous way is planned in the point-to-point communication mode.

8.2.2.1. *Architecture*

A JMS application (Figure 8.3) includes the following elements:

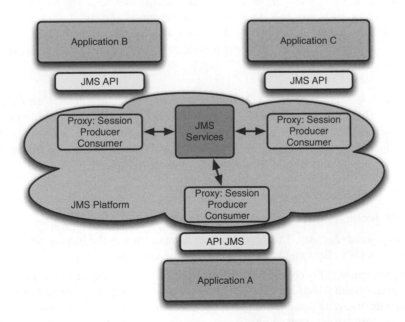

Figure 8.3. *Structure of a JMS application*

– messages. They are objects that can transmit information between different JMS customers. Messages can be in the form of structured text, Java objects or binary format, etc;

– messaging system (*JMS Provider*). The messaging system consists of two elements:

- a basic service that implements the abstractions of programming model as well as administrative and control features,

- a library of functions to develop user applications,

– JMS clients. They are programs written in Java that send or receive messages according to the protocols specified by API JMS.

JMS specification proposes two modes of communication (*Messaging Domains*), point-to-point communication and multipoint communication:

– point-to-point communication. It is based on queues which can contain messages. A client (producer) sends a message to a specific queue where it is taken out by a single consumer only. The consumption of a message can be explicitly made by the consumer or through a monitoring procedure pre-registered by the consumer. A receipt generated by the system or directly by the customer confirms the consumption of the message. The message remains in the queue until it is consumed, or until the expiration of a timer;

– Multipoint communication. It uses the publish/subscribe model. A producer client sends a message addressed to a predetermined object (*Topic*). Customers who were previously subscribed to this *Topic* receive the corresponding message. The consumption of (implicit/explicit) messages is identical to that of point-to-point communication.

The latest version of JMS (JMS 1.1) introduces the concept of *Destination* to represent either a message queue or a *Topic* which can handle both modes of communication in the same way by the client. Transaction mode is possible in two communication modes. In addition API allows *QoS* options to manage temporary or permanent subscriptions and the guarantee of message delivery.

8.2.2.2. *Operational principles*

JMS specification defines a number of abstractions:

– *ConnectionFactory*. This object creates a connection to the messaging system. It is used by a JMS client and includes configuration settings.

– *Connection*. The connection object represents an active connection with the messaging system which allows a JMS client to interact with the system. It authorizes the creation of several sessions. Initially, a connection is in stopped state and can receive messages only after it starts. A connection consumes resource and useless connections should be closed when it is no longer necessary.

– *Destination*. The objective is to establish communication between two JMS clients, Destination object indicates the destination of messages for a producer and the source of expected messages for consumer. It is a queue of messages in the point-to-point communication model and a topic in the publish/subscribe model. It encapsulates different address formats of messaging systems.

– *Session*. In this object, there are contextual information (mono-thread) capable of transmitting and receiving messages.

– *MessageProducer*. This object is created by a session object and is used for sending messages to a *Destination* object.

– *MessageConsumer*. This object is also created by a session object and is used to receive messages left on a *Destination* object.

Table 8.1 summarizes objects according to communication model.

Interface	Peer-to-Peer Communication	Publishing/Subscribing Communication
ConnectionFactory	QueueConnectionFactory	TopicConnectionFactory
Connection	QueueConnection	TopicConnection
Session	QueueSession	TopicSession
MessageProducer	QueueSender	TopicPublisher
MessageConsumer	QueueReceiver	TopicSubscriber

Table 8.1. *Summary*

A JMS client works as follows:

1) it searches a *ConnectionFactory* object in a directory using API JNDI (*Java Naming and Directory Interface*);

2) with *ConnectionFactory* object, it creates a connection and obtains a *Connection* object;

3) with *Connection* object, it creates one or more JMS sessions and obtains Session objects;

4) it searchs in the directory for one or more *Destination* objects;

5) with the help of *Session* object and *Destination* objects, it will create *MessageProducer* objects and *MessageConsumer* objects to send and receive messages:

a) construction of *MessageProducer* objects and *MessageConsumer* objects: a client sends messages to a destination using a *MessageProducer* object. This object is created by the CreateProducer method of *Session* object with a *Destination* object as parameter. If the destination is not specified, each message sending must be passed with a *Destination* object (as parameter of Send method). The delivery mode, priority and lifetime can be specified by the customer for all messages sent by a *MessageProducer* object, or be specified for each message. *MessageConsumer* object allows a client to receive messages. It is created by calling `CreateConsumer` method and *Session* object and by passing a *Destination* object as parameter. We can add a message selector to filter messages for consumption. In JMS, there is a synchronous consumption model (a client invokes the `Receive` method and the *MessageConsumer* object to receive the next message, it is the consumption mode *Pull*) and asynchronous (a client registers a prior object that implements the *MessageListener* class in the *MessageConsumer* object, calling the `onMessage` method on this object delivers messages when they arrive, it is the *Push* consumption mode);

b) creation of *Destination* objects and *Message* objects;

c) transaction support (it allows the creation of groups of received and emitted transactions which can be validated or not);

d) scheduling of received and sent messages;

e) management of acquittal messages.

A JMS message consists of 3 parts, header (used for identification and routing), properties (standard optional fields, or application-specifics, or messaging system), body (JMS defines different types of message's body compatible with various messaging systems).

The header contains the following fields:

– *JMSDestination* contains the message destination, following the method of sending messages according to specified *Destination* object;

– *JMSDeliveryMode* indicates the delivery mode of the message (persistent or not). It is specified by the method of sending messages depending on defined parameters;

– *JMSMessageID* is an identifier that characterizes the message in a unique way. The transmitter can examine it after sending the message;

– *JMSTimeStamp* represents the time of consideration of the message by the messaging system;

– *JMSReplyTo* corresponds to *Destination* to which the client can possibly send a reply;

– *JMSExpiration* contains the sum of the current time and the lifetime of a message (TTL). A message not delivered before the expiry date is destroyed without notice.

– *JMSCorrelationId, JMSPriority, JMSRedelivered, JMSType*, etc.

The properties are of type *String* and can take the following values: `null`, `boolean`, `byte`, `short`, `int`, `long`, `float`, `double`, `String`. They allow a customer to select messages based on criteria. A client can set filters on reception through *MessageConsumer* object, using a string. The syntax of this string is a subset of SQL language.

The body may be one of the following:

– *StreamMessage* contains a series of values of Java primitive types. It is sequentially filled and read;

– *MapMessage* includes a set of name-value pairs;

– *TextMessage* contains a string;

– *ObjectMessage* includes a "serializable" Java object (serialization mechanism of Java objects);

– *BytesMessage* consists of a sequence of bytes. It is used to encode a compatible message with an existing application.

8.2.2.3. *Conclusion*

JMS defines a protocol for exchanging messages between producers and consumers, but provides no guidance on the implementation of the message service which is proprietary. A JMS application is assumed to be independent of a particular JMS platform since it is based on API to access the message service, and hence this should allow portability. However this portability is reduced because the administration functions are proprietary. Furthermore, the interoperability between two JMS clients located on two different platforms is not guaranteed because the manner in which the two platforms function may be different. On the market for JMS platforms, we have for example JORAM [OW2 09] of *ObjectWeb* solution that support this function. This solution is presented in section 9.6.2. They are not considered as essential JMS functions to implement a JMS platform such as: configuration, message service administration, security (integrity and confidentiality of messages) and some parameters of quality of service. Some of these features are present in existing platforms on the market and are proprietary implementations.

8.2.3. *XMPP*

XMPP (formerly *Jabber*[1]) or *EXtensible Messaging and Presence Protocol* is an open standard protocol of instant messaging developed by Jeremie Miller in 1999 then by the *Open Source Jabber* community in the following years. It was ratified by IETF in 2004. It is based on a client-server architecture (Figure 8.4) which enables decentralized exchanges of messages (instant or not instant) between clients in open XML format.

One of the advantages of the XMPP protocol lies in its separation into two different parts:

– Basic protocol on one hand

It contains basic concepts for operating a *Jabber* infrastructure. It is defined by the following RFCs:

---

1 *Jabber* is an instant messaging system based on the XMPP protocol.

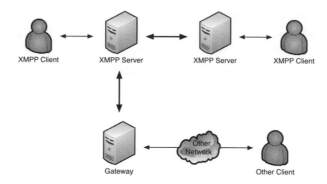

Figure 8.4. *XMPP architecture*

– RFC 3920 [SAI 04a]. It is the heart of XMPP and it describes client-server messages that use two XML streams. The tags are as follows: `<presence/>`, `<message/>` and `<iq/>` (*Info/Query*). A connection is authenticated using SASL (*Simple Authentication and Security Layer*) and encrypted by TLS (*Transport Layer Security*),

– RFC 3921 [SAI 04b]. This RFC describes the most common application that uses XMPP, instant messaging,

– RFC 3922 [SAI 04c]. This is the description of *Common Presence and Instant Messaging* (CPIM) specifications in XMPP,

– RFC 3923 [SAI 04d]. It describes the end to end encryption of XMPP messages by using S/MIME.

Theoretically, such an infrastructure can not function without a full application of these protocols.

– XEPs (*XMPP Extension Proposal*) on the other hand.

These are proposals for new features of the protocol. Servers or clients are not obliged to adopt these extensions because they may block some functionalities between two users. XEP [XMP 08] are continually created, revised or improved as part of XMPP *Standards Foundation* (XSF) formerly known as *Jabber Enhancement Proposals* (JEP).

8.2.3.1. *Principles*

The data is encoded in XML in *Stanzas* which are small packets of information. These *Stanzas* are transmitted through a TCP connection between client and server, or between server and server. The communicating entities are identified by an address which has a syntactic form similar to that of email. XMPP, such as email, is operated by the connection of servers, where each server is responsible for one or more areas. Clients send messages to servers, which are responsible for delivering to the next server. Addresses (JID for *Jabber Identifier*), besides the classical `user@domaine`, can contain a resource, indicated after the slash. A *Stream* is an XML `<stream>` element which includes one or more *Stanzas*. There are many types of *Stanzas* besides error messages. Figure 8.5 illustrates classic example of an exchange (*C* is the client and *S* is the server), with *Stanzas* of message type:

```
C:    <message from='toto@example.com'
                to='titi@example.net'
                xml:lang='fr'>
        <body>An example</body>
      </message>
S:    <message from='titi@example.net'
                to='toto@example.com'
                xml:lang='fr'>
        <body>Answer to example </body>
      </message>
```

Figure 8.5. *Example of an XMPP exchange between client and server*

8.2.3.2. *Security*

XMPP offers several levels of security at the protocol level. It has two security protocols, first TLS (*Transport Layer Security*) for encryption of the channel, then SASL (*Simple Authentification and Security Layer*) for a strong authentication. The negotiation is performed in two phases. There is an opening of a *Stream* for TLS negotiation when this negotiation is successful, a new *Stream* is opened for SASL negotiation. If the application can pass through these two steps, it can open *Streams* for its own operation.

Identity in XMPP is much more stronger than in other systems such as email. Users must authenticate to their home server and messages from users can not be usurped by simply replacing headers, as it can be done with emails. This eliminates or reduces the spam problem. An additional identity verification can be obtained by asking clients to have valid security certificates in order to confirm their identity. In addition, the association of hosts can be easily and tightly controlled by limiting access to a whitelist of a number of participants.

The encryption is done between both client and server and between servers. These two types of encryption are optional in XMPP, but servers can be configured to only accept encrypted connections. Once the connection is encrypted all messages sent between client and server are also encrypted. In addition, for the server-server exchange, administrators can choose to associate with any other open XMPP server, with a set of XMPP servers, or with any other server. This feature of the protocol allows fine control of server clustering for different applications.

8.2.3.2.1. TLS

TLS[2] is described by RFC 2246 and is the successor of SSL. It is standardized by IETF. It operates in client-server mode and has four security objectives:

– server authentication;

– confidentiality of exchanged data (encrypted session);

– integrity of exchanged data;

– option: authentication or strong authentication of client with the use of a digital certificate.

The operation is as follows:

– negotiation of cryptographic algorithms and compression;

– exchange of certificates which makes it possible to calculate a common secret on each side;

– use of common secret to extract cryptographic keys from TLS session.

XMPP use TLS to ensure confidentiality and integrity of data with a STARTTLS extension which is inspired from extensions for IMAP and POP3 protocols. An administrator of a given area can use TLS for client-server or server-server communication or both. Customers should use TLS to guarantee the stream before starting a SASL negotiation, and servers should use TLS between two domains to ensure the security of communications from server to server. TLS protocol respects fourteen rules which were detailed in RFC 3920. Here is an example of a rule: if the initiating entity chooses to use the TLS protocol, TLS negotiation must be completed before SASL negotiation; this order of negotiation is necessary to protect the information of SASL authentication sent during the negotiation. It is also possible to use an external SASL mechanism on a previously planned certificate during TLS negotiation.

8.2.3.2.2. SASL

SASL means "simple authentication and security layer". This protocol is also standardized by IETF. It is described by RFC 2222 (now replaced by RFC 4422). It defines an identification and authentication mechanism between a client and a server through the introduction of a security layer between the protocol and the connection (in the sense of IP). The created layer can also protect the exchange according to the protocol. Authentication mechanisms are separated applications that can use any mechanism supporting SASL by any application supported by SASL. SASL protocol respects ten rules detailed in RFC 3920. Example of a rule: if the receiving entity supports a SASL negotiation, it must have one or more authentication mechanisms

---

2 *Transport Layer Security.*

in a `<mechanism/>` element in response to opening *Tag* of the received stream of initiating entity.

8.2.3.3. *Presence mechanism*

XMPP in its basic version provides a mechanism to view the presence and status of nodes using this protocol. The client, when connected, can send its presence information (by `<presence>` element) either directly to another client or to a server which will distribute them to clients who are authorized to know that information. The source code 8.6 shows an example of notification of the presence of a client to a server.

```
C:    <presence from='toto@example.com'>
          <show>online</show>
          <status>ready</status>
          <priority>5</priority>
      </presence>
```

Figure 8.6. *Example of a presence notification*

The server also sends to the client the presence information of its contacts, either directly if the information is already in its possession, or by broadcasting previously a *Presence* packet of type *Probe* to these same contacts. The contacts of a client are grouped into lists, called *Roster*. A customer must subscribe to a contact, i.e. to be approved by the contact, before they can be warned of his presence.

8.2.3.4. *Extensions*

XSF (*XMPP Standards Foundation*) publishes official extensions to XMPP protocol. These are more than two hundred and fifty. Some of them are listed below:

– XEP-0118 - *User Tune* defines a communication protocol allowing a client to inform those contacts about the music he plays. This information can be viewed as a sort of "widespread presence";

– XEP-0136 - *Message Archiving* allows the archiving of messages on the server, either at the request of client, or automatically by the server;

– XEP-0166 - *Jingle* is a multimedia extension of XMPP, which makes it possible to extend the stream to any binary content as soon as possible: Internet telephony, video conference, etc;

– XEP-0239 - *Binary XMPP* provides encoding of data in binary, and is more efficient than XML. This specification defines XMPP in binary.

8.2.3.5. *Other uses of XMPP*

Instant messaging is just one aspect of XMPP protocol, which is primarily a means of circulating information as an XML stream. We can imagine many other possible applications using extensions proposed in the framework of XEP. Some of the extensions are currently implemented, especially the ones related to basic functions of instant messaging, but many of them are still in a draft form:

– Notifications. One of the most convenient features of instantaneous messaging is to be notified about new event. Possibilities and ideas are numerous:

    - email, arrival of a new message (ILE, JMC, IMN);

    - typing a form (to monitor the typing of an entry in a wiki, a post or comment on a blog, a participation in a forum, etc.);

    - News through the monitoring of *RSS/Atom* (JRS, *pyrss*, *rss2jabber*, *JabRSS*, *janchor*, *Pubsub.com*);

    - personal or group calendar (reminder and appointment alarm);

    - repository monitoring (CVS/SVN), etc.

This type of task is usually performed by robots responsible for monitoring and sending notifications.

– Groupware. XMPP is a communication protocol, and is ideal for remote collaboration. This idea is mostly implemented at the client level: client should be able to accept specific exchanged data.

    - documentation and file sharing;

    - voIP (Jingle, Jabbin, etc.);

    - whiteboard (Coccinella, Inkboard/Inkscape);

    - working together on a document;

    - Co-browsing (Jybe);

    - Record of discussion and notetaking.

– Presence. It corresponds to information that lets you know about the availability state of a *Jabber* client, its location and what it is doing.

    - geo-localization (JEP 112, *Talk Maps*, TRAKM8 which locates vehicles by XMPP);

    - system administration: monitoring the uptime of a remote machine (via a robot);

    - display the presence of a user on a web page (*Edgar the Bot*), etc.

– system administration: surveillance, as well as remote control.

    - remote monitoring of a machine: *Uptime*, bandwidth, *Logs*, attacks, *Cron* tasks, exceeded limits alerts, *Pings*, memory status, network load and CPU, network activity, etc. Once again, we are in the notification domain.

    - remote controls: Updates, *Backup*, query of *Logs*, *Backdoors*, etc.

– various

    - Online games;

    - Updating your blog;

- Consulting search engines, weather or other services;
- Online bookmarks.

This brief overview shows the variety of applications that can be implemented using the XMPP protocol.

8.3. Service-oriented middleware

Service-oriented middlewares are nowadays very popular for the deployment of integrated enterprise information systems. As we have observed in Chapter 7, the service concept has been popularized since the early 2000s by the emergence of *Web* services. The various vendors in the market (Microsoft, SUN, IBM Oracle, BEA, Sybase, etc.) offer different solutions.

In this section, we will examine OSGi platform which is an effort to standardize the concept of services in Java. Several implementations exist, for example IBM Websphere and IDE Eclipse, which offer a fairly widespread implementation. We also present the OSGi specification, its main characteristics and a description of UPnP standard which is often associated with OSGi when the service integration gateway has to manage auto-configurable hardware.

8.3.1. *OSGi*

OSGi alliance [SUN 08] is a standard-setting organization founded in March 1999 by a dozen companies, including Ericsson, IBM, Oracle, Sun Microsystems, etc. Its objective is to propose a set of specifications for gateway software. These softwares are implemented in Java and installed between an external network (e.g., Internet) and a LAN (a domestic network, a sensor network or a network of objects linked by their RFID tags as an example). The central part of this specification is a *Framework* which defines a lifecycle model for applications which are independent of that of Java virtual machine and a mechanism for service registration.

8.3.1.1. *Operational principles*

OSGi platform (illustrated in Figure 8.7) is a platform for deploying and running Java applications that run within the same JVM. An application that is attached to a service provider can possibly work with other providers' applications, hosted on the platform. The platform is generally represented as an overlay to *Java Runtime Environment* which allows dynamic relationships between deployed applications. The characteristics of OSGi include the following:

– modular and extensible (management of dependencies between modules; flexible model that facilitates adding new features);

– dynamic (deployment of new services without platform interruption; dynamic code update);

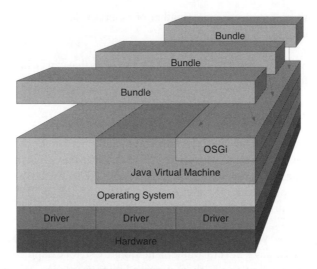

Figure 8.7. *OSGi architecture*

– configurable (each publisher/host only deploys the services they need);

– Tele-administration (Remote Administration; remote deployment of *Bundles*).

The key entities of OSGi are:

– *Services*.

A service is a Java interface which performs a particular feature. This service is programmed in a class that implements this interface. A service is registered to the naming service of *Framework*, with a set of properties of type <name, value> which characterizes it. *BundleContext* provides an interface to *Bundle*. It allows *Bundle* to register a service in the naming service of *Framework* and is to be kept informed of the presence of new services in the system.

– *Bundles*.

They are both functional and deployment units that contain services. A *Bundle* is the delivery and installation unit of Java *Packages*. It is packaged in a ".jar" file which contains the service interfaces and their implementations, a service activation class and a *Manifest* file (metadata). *Manifest* contains the required information to *Framework*, so that it can manage the execution of services of *Bundles*.

– contexts of *Bundle*.

They are the execution environments of *Packages* in the structure.

Based on this *Framework*, many OSGi layers, API and services are defined:

– Log or Data Logging;

– Configuration and Preference Management;

- HTTP service (by executing Servlets);

- syntactic analysis XML, Device Access;

- Package Admin;

- Permission Admin;

- Start Level;

- User Admin;

- IO Connector; IO = Input Output;

- Wire Admin;

- JINI, UPnP Exporter;

- Application Tracking;

- Signed Bundles;

- Declarative Services;

- Power Management;

- Device Management;

- Security Policies;

- Diagnostic/Monitoring and Framework Layering.

8.3.1.2. *Packaging and deploying of applications*

A Java application usually includes many JAR files located in different directories. The paths to these files are listed in the CLASSPATH environment variable. Besides the issues of localizing classes and the risk of errors in the definitions of paths, this approach raises a problem of version incompatibility between classes in operating environment and those in compiling environment.

In OSGi, applications are deployed as packaging and delivery units (*Bundles*). The lifecycle of *Bundles* is as follows:

- installation. Downloading and storage of JAR file of *Bundle* in file system of platform (*Bundle Cache*);

- dependency resolution. Dependencies of *Bundles* installed on the platform are resolved from *Packages* exported by all *Bundles*. Such information about exported *Packages* are listed in *Export-Package*, but they are activated only when *Bundle* seeks them. When several *Bundles* export the same *Package*, only the one having the highest version number and the smallest identifier is exported;

- activation. The platform instantiates one object of the class given by *Bundle-Activator* declared in the *Manifest* file of the *Bundle*. This class must implement *BundleActivator* interface which governs the lifecycle of a *Bundle*. start(BundleContext) method is invoked with a parameter that represents the context of *Bundle* (the platform). This method can register services, search for other services, and start *Threads*;

– stop. *Stop(BundleContext)* method unregisters provided services, releases used services and stops started *Threads*;

– update. Shutdown, reinstall, resolution and reactivation of a bundle without interrupting other services of the platform;

– uninstallation. Delete JAR file in *Bundle Cache*. However classes are still loaded in memory, while bundles that depend on them stay active.

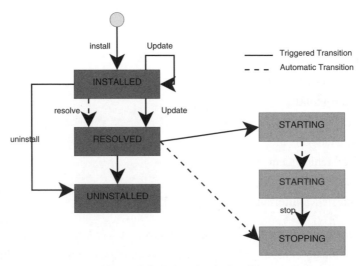

Figure 8.8. *Lifecycle of a bundle*

An OSGi *Bundle* can deliver native libraries, which are described in *Manifest* through `Bundle-NativeCode` entry. This entry describes the path in JAR file, target CPU and OS are required.

OSGi follows the paradigm of dynamic service-oriented programming. This paradigm considers that any service with a required contract is "substitutable" to another and the choice among the list of services with a required contract is decided as late as possible at runtime. Client retrieves this list by querying a register of services and providers register their services with the registry, involving a qualified contract. Moreover, this paradigm is dynamic, it considers that necessary services to a client may appear and disappear at any time. Services are the means for *Bundles* to cooperate with each other. The service contract consists of one or more Java interfaces described by a set of mandatory and optional properties. The interfaces are used for syntactic negotiation, while the properties are used for Quality of Service (QoS) negotiation.

8.3.1.3. *Security*

One of the objectives of OSGi is to run applications from different sources under strict supervision of a management system. A comprehensive security model

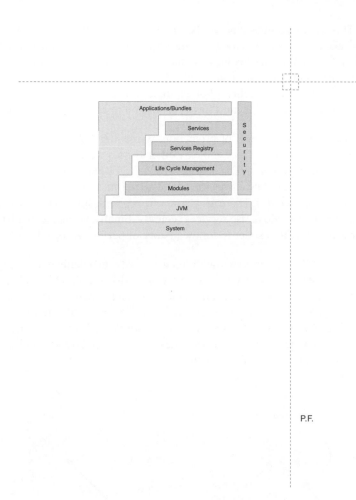

Figure 8.9. *OSGi architecture and security model*

(Figure 8.9), present in all parts of the system, is then necessary. OSGi specifications use the following mechanisms:

– Java security;

– reducing the visibility of *Bundles*'s content;

– management of communication links between *Bundles*;

– secure deployment of *Bundles*.

Java security model provides the concept of permissions that protect the resources of specific actions. For example, files can be protected by action permissions. Each *Bundle* has a set of permissions that are subject to some conditions (Java signature, path, etc.). This set of permissions can be changed on the fly and new permits are immediately taken into account. However, the allocation of these permissions can also be done earlier, for example, during the installation phase. When a class wants to protect a resource, it requests *Security Manager* to check if the class in question has permission to perform an action on this resource. For example, if *A* calls *B* and *B* has access to a protected resource, both *A* and *B* must have permission to access the resource.

OSGi adds an additional level of protection by making *Packages* visible only in the *Bundle*. These *Packages* are accessible by other classes within the *Bundle*, but they are not visible to other *Bundles*. The sharing of *Packages* between *Bundle* may pave way for a malicious attack. OSGi specifies delicate permissions of packages to limit exports and imports in *Bundles*. There are also services permissions. These permissions provide the ability to provide or use services securely. Finally, a *Log* mechanism keeps track of access to services in order to follow history of performed operations.

Version 4 of OSGi incorporates additional security mechanisms for deploying *Bundles*. In this operation, we try to ensure the integrity and confidentiality of data and to authenticate the sender. Initially, the supplier obtains a pair of private/public key from a trusted organization. The client retrieves a public key certificate from the same organization. Provider signs and authenticates its *Bundle* with the help of its private key. The client downloads the *Bundle*, validates it with its public key and installs it. The mechanism is illustrated in Figure 8.10.

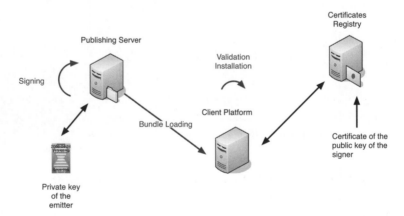

Figure 8.10. *Secure deployment of OSGi*

8.3.1.4. *Implementations*

There are a dozen implementations in the market, proprietary and *Open Source*. Some implement all the specifications given by the alliance, while others only partially. In addition, each one of them has their own features such as the presence of a GUI or a single console to control *Framework*.

Oscar and Felix are the most popular and the simplest to implement in *"Open Source* world". *Oscar* is an implementation whose advantage is the ability to operate on equipments with limited memory resources. In fact, it can be installed on small devices that only have a small memory footprint. It has a console for installation and control of *Bundles*. Unfortunately the provided *Framework* only respects recommendations of OSGi *Release* 3. The application is no longer in development stage, and the team in charge of the project has decided to participate in development of another platform (*Felix*). This one must comply with specifications of *Release* 4.

Knopflerfish is an OSGi and *Open Source* implementation which, by default, uses a graphical interface with all the required functionality to have an effective management of *Bundles*. It partially implements in its version 2 recommendations of *Release* 4 of OSGi specification. Today, *Knopflerfish* is the most successful *Open Source* OSGi platform. It is in the core of *Eclipse* development environment which ensures a wide dissemination.

Concierge is very suitable for "Embedded Systems World" with very low resources, in particular mobile phones, as their memory does not exceed 80 KB. This distribution is not based on a complete JVM but profiles of J2ME. The distribution is *Open Source*.

8.3.2. *UPnP*

UPnP[3] Forum, formed in 1999, is an open industry consortium. The Forum's goals are to define standards simplifying network implementation of communicating equipments in home and corporate environments (SOHO: *Small Office Home Office*). The covered areas are micro-computer/office (printers, scanners, APN, etc.), consumer electronics (DVD, TV, radio, *Media-Center*, etc.), communication (telephones, ADSL router, etc.), home automation (burglar alarm, heating control, shutters, etc.) or even home appliances (washing machine, refrigerator, etc.).

UPnP brings together a set of network protocols which enable devices to progressively connect and to simplify the implementation of networks in home environment (data sharing, communications and entertainment) and corporate environment. UPnP fulfills this mission by defining and using device control protocols built upon open, Internet-based communication standards (TCP/IP, UDP and HTTP).

UPnP allows communication between any two devices through a network control device. The main highlighted features in this architecture are:

– support for several means of communication (telephone lines, powerline, Ethernet, Infrared, Wi-Fi, Bluetooth and FireWire);

---

3 UPnP: *Universal Plug and Play.*

– independence between devices and means of communication;

– no device driver is used, only standard protocols are.

UPnP Forum has published a first version of required network protocols (UPnP DA) and a number of standard definitions of device and their associated services. UPnP *Device Architecture* (DA) defines protocols for the constitution of spontaneous network of device management: (*i*) dynamic detection and removal of devices; (*ii*) descriptions of devices and services they provide; (*iii*) utilization through control points (PDA, TV, RF remote, etc.) of provided services; (*iv*) notification of changes of state variables associated with services.

UPnP architecture supports "zero configuration" mode: each device can dynamically join a network, obtain an IP address, announce its name, provide information about its functionality on demand and discover the presence and functionalities of other devices on the network. Utilization of DHCP and DNS are optional. Devices can dynamically leave the network without leaving undesirable information after their exit.

Various protocols proposed in UPnP correspond to different phases of lifecycle of an equipment that is connected to a network.

8.3.2.1. *Discovery*

Equipped with an IP address, the first step in a UPnP network is the discovery of resources. When a device is added to a network, the discovery protocol SSDP[4] of UPnP allows this device to register its services to control points of the network. Similarly, when a control point is added to the network, the discovery protocol enables the control point to look for interesting devices on the network. The fundamental exchange in these 2 cases is a discovery message containing some essential information about the device or its services (type, identifier, a pointer on more information, etc.).

8.3.2.2. *Description*

After a control point has discovered a device, it knows very little about the device. To learn more, or to interact with it, it must find the description from URL provided by the device during its discovery message. UPnP description of a device is written in XML and contains information such as manufacturer's model name, serial number, URL of seller's website, etc. The description also includes a list of all embedded devices or services, such as URL to monitor events and presentation. For each service, the description includes a list of commands or actions to which the service responds, and parameters or arguments for each action. The description of a service also includes a list of variables. These variables describe the state of service during runtime and are described in terms of data type, characteristics of events, etc.

---

4 *Simple Device Discovery Protocol.*

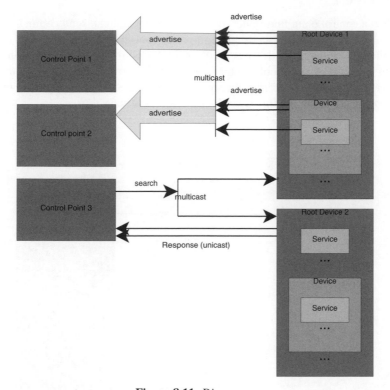

Figure 8.11. *Discovery*

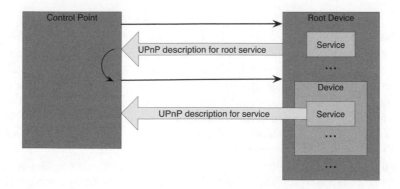

Figure 8.12. *Description*

8.3.2.3. *Control*

When a control point finds a description of a device, it can send commands to the device's service. For this, it sends an appropriate control message to control URL for the service (provided in the description of device). Control messages are expressed in XML by using SOAP protocol. As for function calls, in response to control message, the service returns the expected values in action. The effects of the action are modeled by changes in the variables that describe the state of service.

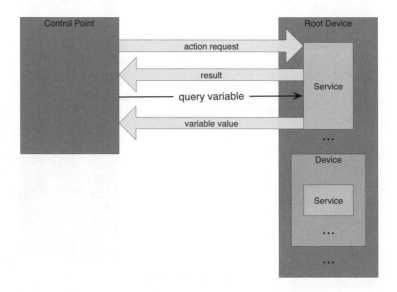

Figure 8.13. *Control*

8.3.2.4. *Event notification*

UPnP description of a service includes a list of actions to which the service responds and a list of variables that models the state of service at runtime. The service publishes updates when these variables change and a control point must subscribe to receive this information. The service publishes updates by sending event messages that contain the names of one or more state variables, and the current value of these variables. These messages are expressed in XML and formatted using *General Event Notification Architecture* (GENA). A special initial event message is sent when a control point subscribes first time. This message contains the names and values of all variables involved in the event, and allows the subscriber to initialize its state model of service. To support scenarios with multiple control points, the events are designed to keep all control points equally informed about the effects of any action. Thus all event messages are sent to all subscribers irrespective of the reason for which the state variable is changed.

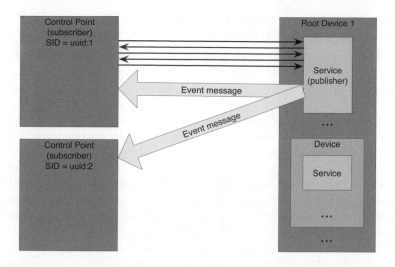

Figure 8.14. *Event notification*

8.3.2.5. *Presentation*

If a device has a URL for presentation, the control point can retrieve a page from that URL, load the page on its browser, and depending on the features of the page, it allows a user to control the device and/or see its state.

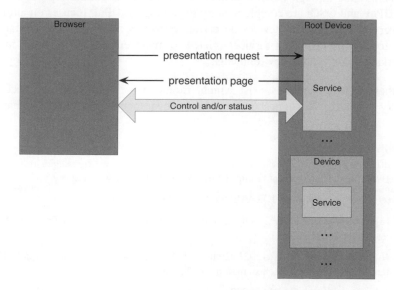

Figure 8.15. *Presentation*

To function, UPnP uses a variant of HTTP on UDP (HTTPU and HTTPMU, *Unicast* and *Multicast*), whose specification is subject of an *Internet Draft* which expired in 2001.

8.4. Conclusion

We have recollected the standards proposed by the EPCglobal consortium to promote a network for the Internet of Things which allows the deployment of distributed applications using RFID tags. The standards are used by majority of solutions which we will study in the next chapter. The standardization efforts are aimed at promoting the interoperability between different vendors/subscribers of EPCglobal network. However, they provide only limited tools for integrating to an existing information system, so that different business applications can derive full advantage of this new architecture. To achieve this, major vendors employ complementary technology choices based on standards which are not necessarily related to the problem of the Internet of Things, but are rather related to the field of enterprise application integration. In this chapter, we presented some of the standards related to message-oriented communication middleware. In particular, JMS standard is very popular in the Java world and is implanted in almost all solutions based on this environment. While the XMPP standard ratified by IETF is not specific to a platform and offers better prospects for interoperability.

In the field of service-oriented middleware, many solutions are available. We chose to illustrate this through OSGi specification which has a certain reputation in this field. UPnP and OSGi are complementary technologies. UPnP focuses on automated discovery of services and devices on private networks based on Internet standards. OSGi focuses on providing services related to management of bandwidth in home networks, cars or in other environments.

Other industry solutions exist which, though widespread, are not necessarily industry standards. We suggest that readers discover them in the following chapter.

8.5. Bibliography

[OMG 09] Object Management Group, CORBA, 2009, http://www.omg.org.

[OW2 09] OW2 Consortium, JORAM, 2009, http://joram.ow2.org.

[SAI 04a] SAINT-ANDRE P., RFC 3920 - Extensible Messaging and Presence Protocol (XMPP): Core, memo, IETF, 2004.

[SAI 04b] SAINT-ANDRE P., RFC 3921 - Extensible Messaging and Presence Protocol (XMPP): Instant Messaging and Presence, memo, IETF, 2004.

[SAI 04c] SAINT-ANDRE P., RFC 3922 - Mapping the Extensible Messaging and Presence Protocol (XMPP) to Common Presence and Instant Messaging (CPIM), memo, IETF, 2004.

[SAI 04d] SAINT-ANDRE P., RFC 3923 - End-to-End Signing and Object Encryption for the Extensible Messaging and Presence Protocol (XMPP), memo, IETF, 2004.

[SUN 08] OSGi Alliance, http://www.osgi.org, 2008.

[SUN 09] SUN, JMS, 2009, http://java.sun.com/products/jms.

[XMP 08] XMPP Standards Foundation, XMPP Extensions, 2008, http://xmpp.org/extensions.

Chapter 9

Middleware for the Internet of Things: Some Solutions

In this chapter, we will consider in detail some of the available solutions. As we write this book the middlewares, which are discussed here, are commercial products offered by major industrial participants in the field, or solutions proposed by more or less extensive consortia, or more academic solutions developed by academic groups which are often smaller but have innovative features. Initially, we will consider the solutions of major editors of professional markets, in particular the solution proposed by Sun with its software suite: Java SUN RFID System. In the industrial field, competitors have been in a hurry to rush to propose solutions to effectively integrate the RFID technology in their information systems:

– IBM proposes a solution IBM *Websphere* RFID Suite, which is based on the *Device* and *Enterprise* editions of *Websphere*;

– BEA with the purchase of one of the pioneers in the domain, *ConnecTerra* and its *ConnecTerra RFTagAware*;

– *Oracle* and its *Oracle Sensor-Based Services*, which integrates informations acquired on sensors or RFID tags in the *SI* of an enterprise. This solution is an RFID driver kit and is integrated into the support of *Oracle e-Business Suite* and of the *Oracle Application Server*;

– Sellers of integrated, execution and optimisation solutions on the supply chain such as Manhattan Associates (900 major customers worldwide), or SAP or *Savi Technology SmartChain* (first RFID middleware provider of DoD);

Chapter written by Yann Iagolnitzer, Patrice Krzanik and Jean-Ferdinand Susini.

– Microsoft, despite a late entry in the market, offers a software suite built on its flagship technology for enterprise: Microsoft.Net RFID *Services Platform*.

Note that all these solutions have adopted the EPCglobal technology. Their editors are participants in the consortium and most of them are active in its promotion and its development. These products are very similar at the architectural level due to the EPCglobal standards specified to facilitate the interoperability in the EPCglobal network. If the global architecture of middlewares remains the same between different competitors, the implementation of different components can differ depending on the choice of implementation technology.

In this chapter, we present the main solutions of SUN, IBM and Microsoft.

9.1. EPCglobal and SUN Java RFID software

Using the standards of EPCglobal, SUN [SUN 09] proposes a set of products for RFID tag management in order to automatically and uniquely identify a very large number of objects. These solutions allow us to track objects, trigger events, and even perform actions on objects themselves. These solutions are primarily focused on the management of supply chains, and enable enterprises to verify the integrity of their stocks and thus make savings, by increasing the efficiency of their operations. As a member of *Auto-ID Center's Technology Board* and of *MIT Auto-ID Center's Software Action Group*, SUN directs the efforts of the industry to RFID/EPC standards. SUN participates and leads these efforts as a member of EPCglobal, and provides a dedicated infrastructure based on RFID/EPC standards and solutions to deploy RFID applications for enterprises. Finally, SUN is significantly involved in the deployment of the EPCglobal network infrastructure. This last point is also an opportunity to challenge the relevance of the EPCglobal model by its critics who see in it a provision by enterprises using the EPCglobal network and its services, important information on the activity and the situation of a company *vis-à-vis* its stocks, between private hands.

9.1.1. *Software architecture of SUN Java System RFID*

It is built with the *Java Enterprise System* software and technology [SUN 09], which provides a set of technology components and products to improve the integration and simplify the maintenance. The result is a set of firm infrastructure services with paying annual license (for software, data hosting, maintenance, consultations and trainings).

The solution of SUN proposes a driver set to directly use many RFID devices (some branded readers: *ThingMagic*, *SensorMatic*, *FEIG Electronic*, *Zebra Technologies*, *SAMsys*, *AWID*, *Printronix*, *Symbol*). SUN also proposes the Java System RFID Software Toolkit, designed to simplify the construction of adapters for different readers,

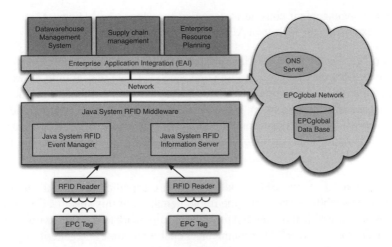

Figure 9.1. *General architecture of the SUN-EPCglobal solution*

and printers or other devices compatible with the event manager to enrich the initial hardware support. It provides a dedicated development environment. SUN also proposes a JVM supporting some RFID readers, which makes it possible to embed on readers themselves higher-level processing (filtering, aggregation, correlation, etc.) specific to the EPCglobal middleware.

Based on its exclusive rights obtained from the EPCglobal consortium, SUN proposes a RFID test center based in Dallas, in Texas, to test solutions on a large scale to help firm users to design and deploy their RFID solutions. This center provides an environment that simulates real conditions (17,000 m^2), with conveyors, pallets, antennas, tags, readers, etc. It is resolutely turned toward mass-market retailing (Wal-Mart, etc.). Finally, SUN uses its *Java Enterprise System* solutions as a base which allows us to use the developments on *Web* services and Java application services.

9.1.2. *Java System RFID event manager*

It is the implantation of a key component of the EPCglobal software stack: the event manager (ALE). It adheres to the standards proposed by EPCglobal and offers additional features designed to facilitate the implantation of large scale systems. According to the specification of EPCglobal, the middleware manages events and information in real-time, generates alerts, and sends information read by a reader to other information systems of the enterprise. The middleware is responsible for processing data of an RFID tag having an EPC code. It provides an interface that allows RFID to be connected to the EPCglobal network. This middleware is compatible with the EPC *Gen* 2 RFID tags as well as with other types of tags that meet the ISO standards and also with other types of sensors.

A center management console is designed to manage each *Event Manager*. The system is extensible through Java API.

Event Manager facilitates the integration of RFID event data with EIS by defining a set of interfaces that manages sending and receiving data in real time. It provides a system of filtering, aggregation and event routing. At each reading stage, the system collects an amount of information, such as the EPC code of an RFID tag, EPC identifier of the reader who has read the tag and a reading timestamp, etc. Information can then be used by filtering and aggregation processes during their propagation in an information system.

The version 1.0 of the EPCglobal middleware specification is built around the concept of extensibility. Therefore, the implementation of this specification primarily defines basic processing modules and provides a framework to develop its own modules to an user enterprise. The RFID event manager enables flexible and generic deployment by trying to preserve aspects of availability, scalability and manageability, and to do this, it is built on a distributed architecture of federated services.

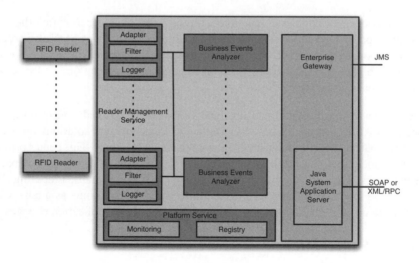

Figure 9.2. *Architecture of the SUN Java Sytem RFID solution*

This architecture is fault tolerant. If a resource is disconnected or damaged, the manager continues to work by redeploying missing software resources. In essence, resilience to failure is due to JINI technology.

The key features of the event manager are as follows:

– adapters allow equipment from different manufacturers to communicate and interact with each other;

– filters are used to filter useful data. Standard implementations of these filters are provided. They smooth out the occurrence of events, manage event changes, and block the consideration of some events;

– event loggers. Standard implementations are provided for writing information in files or in a JMS queue under the form of XML messages by using HTTP or SOAP for example;

– an external interface for other enterprise gateways using data filtered by the event manager.

In order to standardize access interfaces to RFID reader, the solution of SUN proposes several standard protocols:

– *Simple Lightweight* RFID *Reader Protocol* (SLRRP), light version of the SRRP protocol, which was proposed by IETF to exchange data and control information between controllers and RFID readers, when they are connected to an IP network.

– *RFID-Perl*, exchange interface with RFID readers, ensures independence from the heterogeneity of structures.

– JSR-257 specification whose first version was completed in October 2006 and contains optional packages of J2ME for contactless communications.

9.1.3. *Java System RFID information server*

This server provides access to business events generated by the event manager. It serves as an integration layer between the event manager and the existing enterprise information systems (EIS) or applications specific to an user enterprise and provides access to EPCglobal network data to other enterprise applications. In particular, these data include:

– data collected by the event manager through readers and sensors;

– specific data on RFID tags (date of manufacture, etc.);

– information about a product.

Thus, an user enterprise has a system to correlate EPC events with business logic. In addition, the server can be used to implement additional features, such as data reformatting, etc. The server proposes 3 options for point-to-point integration with EIS:

– the J2EE *Connector Architecture* (JCA), to couple EIS to a *Web* service or to a *Web* application;

– sending an asynchronous message by a message-oriented middleware (MOM) using JMS with *Web* services or applications.

To sum up, EIS and the server are considered as *Web* services. They can communicate and interact by using widely deployed standards such as WSDL, UDDI and SOAP.

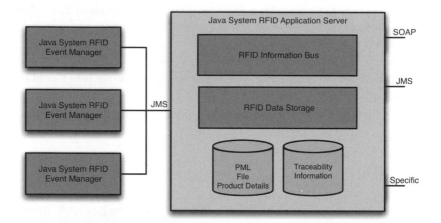

Figure 9.3. *Structure of the Java RFID IS information server*

9.2. .NET and RFID services platform

RFID solutions proposed by Microsoft are primarily based on extensions of products based on .NET platform. Therefore we will quickly present this platform and its specificities in terms of distributed application management before presenting more particularly the RFID *Service platform in detail*.

9.2.1. .NET platform

Today it is a technology commonly used for IT projects based on Microsoft technologies. Microsoft.NET [SIK 09] is a software architecture designed to facilitate the creation, deployment of applications in general but more specifically *Web* applications or *Web* services. It concerns both clients and servers, which will interact with each other using XML.

There are 4 main elements:

– A programming model that takes into account the problems related to deployments of applications (life cycle, version management, security, etc.).

– Development tools whose core is constituted by IDE *Visual Studio.NET*.

– A set of server systems represented by the *Windows Server*, *SQL Server* or *Microsoft Biztalk Server* products which integrate, execute and manage *Web* services and applications.

– A set of client systems represented by workstations under the *Windows* environment, or systems embedded under *Windows CE* or mobile, and equipped with office clients such as *Microsoft Office*.

.NET is a development and execution environment that uses among others the concepts of Virtual Machine (VM), via *Common Language Runtime* (CLR): the source code compilation generates an intermediate object in the MSIL (*Microsoft Intermediate Language*) language and is, independent of any processor architecture and any host operating system. The intermediate object is then compiled on the fly, or in CLR, using a JIT (*Just In Time*) compiler which transforms it into machine code linked to the processor on which it resides. It is then executed. Contrary to Java, for which the intermediate code (*Byte Code*) is related to a single language source, the intermediate code of the .NET platform is common to a set of languages (C#, VisualBasic, C++, etc.). The CLR supporting its implementation contains a set of basic classes related to the management of security, memory, processes and threads.

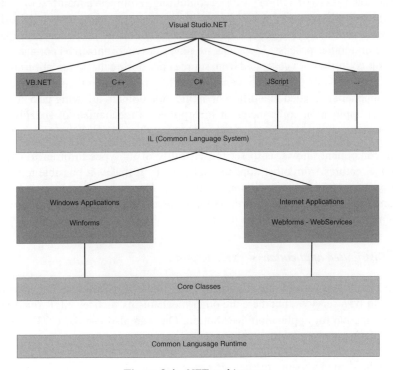

Figure 9.4. *.NET architecture*

9.2.1.1. *Security of .NET*

The deployment of applications is based on the copy in a specific directory, *Global Assemblies Cache* (GAC), or in the user tree, of generated executable files. These executable files (called *Assembly*) are assembled in packages consisting of elements, before being simultaneously deployed. They are usually distributed over several physical files. Using metadata and reflection, all components including their versionsare are self-describing. Any installed application is then automatically associated with files that are part of its assembly.

The security of the code is partially based on the possibility of signing it using a private key in order to identify the person or the organization that has written the code using the associated public key. If the execution in the current directory does not require the signature of the code, any publication in GAC (*Global Assembly Cache*) and release management requires it.

9.2.1.2. *Components of .NET*

The basic block of a .NET application is a component or *Assembly*. This may be an executable (.exe) or a library (.dll) containing resources, a manifest or metadata. Contrary to CORBA or to Java RMI, no other file is required, niether IDL file, nor type library or client/server couple is required to access a component from another language or another process. Metadata are never desynchronized as compared to the executable code, they are generated from source code during the compilation and stored in the executable code. The component is a single deployment unit and a single thread. Each component is loaded in a different application domain allowing it to be isolated from other application components. It incorporates a mechanism of specific version management.

The component allows us to set the visibility of its types in order to expose or hide some features from other application domains. It makes it possible to secure all the resources it contains in a homogeneous way. The manifest contains a list of the types and resources visible outside the component and information about dependencies between the various resources that it uses.

9.2.2. *Distributed applications – .NET Remoting*

.NET provides various mechanisms to distribute applications. The structure is oriented to *Web* services, but there are other mechanisms such as .NET *Remoting* that is more effective for application distribution. One can also use the *COM+* or MOM MSMQ platforms.

With the second version (the new *Microsoft Windows Communication Framework* – Indigo platform) the trend is to unify the development and deployment of distributed applications on .NET. The *.NET Remoting* structure allows the dialog between applications on different application domains. The architecture is modular. The aspects of message transportation, data encoding, remote call are well separated and it is possible to customize each step using specific mechanisms.

There are three types of *.NET Remoting* services:

– services activated by a single-call server (*SingleCall* mode);

– services activated by a shared server (*Singleton* mode);

– services activated by a customer.

.NET Remoting applications interact through a communication channel through which messages flow. We can program the channel type and the way messages are encoded. There are 2 types of standard channels: *HttpChannel* and *TCPChannel*. The system also provides two standard mechanisms for data formatting: *BinaryFormatter* and *SOAPFormatter*.

In order to simplify a maximum creation of a *Web* service, all aspects related to SOAP and WSDL are transparent. Moreover, the *Active Server Pages* mechanism of .NET (ASP.NET) makes it possible on its side (like its competitor Java Server Pages: JSP) to integrate the .NET code in a Web page.

9.2.3. *RFID Service Platform*

It is on the .NET platform that Microsoft has naturally chosen to implement its RFID solution: RFID *Services Platform* [RFI 09]. Microsoft has established a working group: RFID *Council*, is bringing together a number of partners such as Accenture, GlobeRanger or Intermec Technologies to preside over the evolution of the solution. This platform uses the *Microsoft SQL Server* database and the *Microsoft Biztalk Server* integration server.

In addition, Microsoft also wants to incorporate RFID features in business management applications such as *Axapta* 4.0, *Navision* 5.0 and then *Great Plains*. This software aims at making a link between equipment to manage the reception and emission of radio signals from RFID tags, and applications which can interpret these data. It is intended for both enterprises that integrate the RFID management in their information systems and for editors who perform more advanced programs around electronic tags.

The editors also plans the release of a system specifically dedicated to receipt: *Windows XP Embedded for Point of Service* (WEPOS), for retailers which will integrate RFID functions.

9.2.3.1. *Architecture*

The architecture is structured in layers which are described below:

– *Device Service Provider Interface*. This layer is composed of a set of generic API which allows the development of interfaces to integrate RFID equipment in a *Windows* environment. To facilitate the integration, Microsoft provides the platform, specifications and the test software under a form of a SDK.

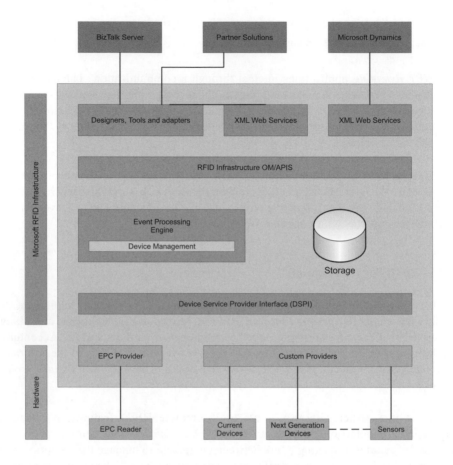

Figure 9.5. *RFID .NET architecture*

– *Engine and Runtime*. This layer facilitates the use of RFID equipment by eliminating the noise and the irrelevance of the raw data provided by tags. This layer allows applications to filter, aggregate and transform RFID data using event managers based on a system of rules. The event processing engine can manage RFID processes irrespective of the types of devices and communication protocols. Its kernel is the event pipeline. It provides the possibility to gather RFID equipment into logic groups. A key element of the event processing engine is the event manager which allows application developers to define a logic for RFID event handling and processing based on the business logic to be implemented. Designed to be flexible and extensible, the event manager automatically applies policies (via business rules) to filter, alert, enrich or transform events.

Another important part of the engine is the management of RFID equipment which allows end users to:

- view the status of an equipment;

- view and manage device configuration;

- access equipment in a secure manner;

- manage addition, deletion, and modification of the name of devices and maintain the consistency in the architecture.

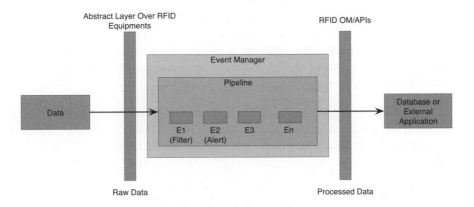

Figure 9.6. *Event manager*

– *Microsoft RFID Infrastructure OM/API*. This layer provides the object model and APIs to design, deploy and manage RFID solutions. It also includes useful tools to manage event processes needed to filter, aggregate and transform data into usable information. The object model (which includes the RFID equipment management, design and deployment processes and event monitoring) provides APIs needed to design and deploy an end-to-end RFID applications and manage it throughout its lifecycle.

– *Designers, Tools & Adapters*. This is a toolset designed to help developers create different types of business processes. For example, *Adapters* helps them to integrate RFID events in real time with *BizTalk Server*.

Two additional tools are provided with *BizTalk* RFID: an administration console called RFID *Manager*, and a tool for writing rules called *Rule Composer tool*. Finally, *Adapters* make it possible to exchange data between *BizTalk* RFID and its users.

9.2.3.2. *Conclusion*

Microsoft provides a platform on which partners can build RFID solutions which will reduce human errors in data collection, reduce inventories and improve the product availability. The Microsoft platform interfaces with existing products such as *SQL Server* and *BizTalk Server* to manage and integrate RFID data, *Visual Studio* and *Web Services Enhancements* (WSE) of *Web* services.

Microsoft BizTalk RFID Mobile is an extension dedicated to mobile platforms (phones, PDAs, etc.) enabling the management of RFID tags, barcodes and other types of embedded sensors.

9.3. IBM Websphere RFID Suite

Like SUN, IBM proposes different test centres in North America, Europe and Asia, to customers for its solutions. To do this, IBM has created a specific division: *Sensor and Actuator* [IBM 09]. This division covers the entire spectrum of RFID solutions and sensors (tags, readers, wireless solutions, middlewares, application-oriented businesses, etc.).

The middleware solution proposed by IBM is based on *Websphere* products and is architected on the base of an enterprise production processes modeling, using the SOA approach. Figure 9.7 shows the main components of this architecture.

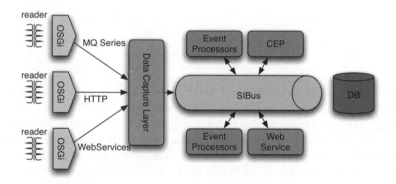

Figure 9.7. *General architecture of the RFID solution of IBM*

9.3.1. *Data capture layer*

Technically, the infrastructure includes OSGi gateways (called *Edge Server*) which directly support RFID readers. These gateways can support different types of readers and send information to information systems of user enterprise. *Edge* servers are generally deployed as close as possible to readers: in warehouses (once again the flagship application of these techniques is the management of supply or production chains). These gateways are responsible for passing read information to the enterprise in "real time". Aggregated RFID events can then be routed to specific servers (called *Premises*) via a message-oriented bus (*MQSeries* or by using *Web* services). The first interface of RFID data sending is a software layer called *Data Capture* whose distributed implementation can be deported to the nearest source of events, i.e. on *Egde* gateways. This component allows us to trace, filter, aggregate and annotate RFID data.

9.3.2. *Premise servers*

Formatted events are placed in a communication channel where they are processed by other components of *Premise* servers. Events transmitted on the communication channel use a format called CBE[1] which is derived from the WSDM[2] *Event Format* (WEF) specification proposed by the OASIS group [WEF 06] in the framework of *Web* services. The message-oriented bus used to route events is generally based on SIBus[3] of the *Websphere* middleware. These technological options facilitate the integration of this architecture in SI based on *Web* services.

Generated events are subject to dedicated processes, which are responsible for analysing received events, establishing correlations based on business-oriented rules and triggering actions or generating new business-oriented events. Processing services connected to SIBus can be relatively simple services (generation of alerts, persistence through a storage on a database, identification, location, etc.). A particular component can manage complex reactions: *Complex Event Processor* (CEP) that captures the business expertise specific to user enterprise by implementing sophisticated processing and reactions based on a set of business-based rules applied by a generic execution engine interpreting rules. Without processing procedures, this system is disabled and the platform is used in its most generic form.

The events running on SIBus are processed by *Task Agents* implanted by *Message Driven Beans* (MDB) using JMS to communicate. This processing chain allows an integration with enterprise servers and in particular BPM solutions proposed by IBM. For this purpose, *Premises* servers propose some business-oriented *Web* services which can be invoked by the BPM environment.

In terms of system administration, all components of the suite are manageable via JMX consoles.

The *IBM Websphere Premise Server* solution provides a framework for facilitating the integration of RFID tags in enterprise information systems. This solution can integrate all types of real time event data (all types of sensors and measurement instruments) and is more generic than EPCglobal solutions, with which it is compatible.

---

1 *Common Base Events.*

2 *Web Service Distributed Management.*

3 *Service Integration Bus.*

The integration is resolutely oriented to SOA architectures of enterprise information systems basing on business *Web* services.

9.4. Singularity

Singularity [SIN 09] is an *Open Source* initiative (the license chosen for the *Singularity* project is the version Apache 2.0) dedicated to the development and the promotion of an RFID software technology in supply chains, EPCglobal network, inventory management, payment solutions etc. *Singularity* is sponsored by *I-Konect*, a service integration company focused on RFID technologies and related technologies, and their applications in industry. *I-Konect* has founded *FirstOpen*, an *Open Source* consortium specialized in solutions based on events ordered by RFIDs and sensors. *Singularity* is the initial project of the *FirstOpen* consortium. The project began on March 31st, 2005 and the first version was born on June 6th, 2006.

The architecture of *Singularity* has two main components: (*i*) *EPC Information Service* (EPCIS) and (*ii*) middleware.

Singularity presents an EPCIS which supports the EPCglobal specification, and allows a perfect integration of information related to EPC code in enterprise. The middleware provides the management of events and RFID readers. The technology of the platform chosen for *Singularity* is Java, since it is present in the infrastructure of most of the enterprises. The main purpose of *Singularity* is to accelerate the development and adoption of RFID solutions. The RFID middleware and EPC-IS provide a platform that reduces input blocking, and a basis that allows enterprises to improve the offer of their products.

In the following sections, we will initially present the middleware and its major components, and then we will present the information server according to the EPC-IS specification. Finally we will discuss some technological aspects related to the implementation of these different components.

9.4.1. *Middleware*

Singularity is a distributed system which enables services to run on different physical or virtual servers. The middleware part consists of three installable major components:

– *Device Manager* (DM). It is a JINI service that individually controls devices on the *Singularity* network such as RFID readers or printers. A *Device Manager* can manage many features and there may be many instances of DM deployed in the network. To start a DM, an appropriate software must be configured and loaded to communicate with devices. It receives configuration information and the software from any CM service which it can find in the network.

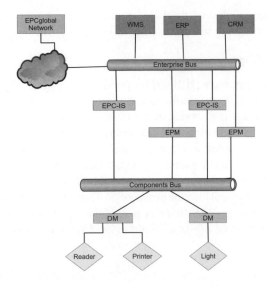

Figure 9.8. *Singularity architecture*

– *Event Process Manager* (EPM). It is a J2EE application deployed as an enterprise archive (EAR). It provides event services at an application level (Application Level Events services according to the EPCglobalALE 1.0 implementation), Complex Event Processing services, process management services, configuration and administration management.

– *Configuration Manager* (CM). CM is a JINI service which can be deployed anywhere in the network. It provides the management and the propagation of configurations for *Device Managers* (DM). It also provides JINI research services as a code server for other services of *Singularity*, such as the code loading for DMs.

When testing, emulators of RFID readers (EM) are proposed and executed under J2SE 5.0. EM is a Java service and not a JINI service.

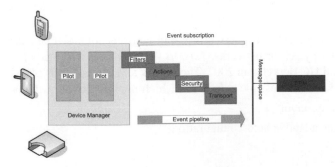

Figure 9.9. *Singularity middleware*

Singularity needs relational databases management systems (RDBMS), usually MySQL, and an application server, (usually JBoss) and Java.

9.4.2. *Hibernate–JBoss*

Singularity uses *Hibernate* [HIB 09] for persistent storage of objects, which allows us to use many RDBMS, *Open Source* or commercial sources without specific code for each of them. *Hibernate* is a service of query and persistence service for objects and relationships. It makes it possible to develop persistent classes according to the object-oriented paradigm, including association, inheritance, polymorphism, composition, and collections. Queries can be expressed in its own portable extension of SQL: HQL[4], both in native SQL and with an object-oriented API. Contrary to other persistence solutions, *Hibernate* does not hide the power of SQL and guarantees business investment in a relational technology. *Hibernate* is a professional *Open Source* project and a fundamental component of *JBoss Enterprise Middleware System* (JEMS) products [JBO 09].

The core of *Hibernate* is *Hibernate* for Java, native APIs and metadata of XML correspondences. One can find:

– *Hibernate Annotations* associates classes with annotations (JSR 175) of JDK 5.0;

– *Hibernate EntityManager* is the Java persistence API for Java SE and Java EE;

– *NHibernate* is the version of *Hibernate* for the .NET platform;

– *Hibernate Tools* are development tools for *Eclipse* and *Ant*;

– *Hibernate Validator* is an API of validation and of data integrity annotation;

– *Hibernate Shards* is a framework of data horizontal partitioning;

– *Hibernate Search*, an integration of *Hibernate* in *Lucene* is to index and acquire data;

– *JBoss Seam* is a framework for JSF, AJAX, EJB 3.0 and Java EE 5.0 applications.

9.5. Middleware for embedded systems

9.5.1. *TinyDB*

It is a system for "data" query processing which is not based on a database but on data collected by sensor networks. *TinyDB* [TIN 09] was designed at Intel-Research Berkeley, and UC Berkeley in 2002. The goal is to meet the needs of sensor networks located in places inaccessible to humans, which implies the following properties:

– energy autonomy;

– dynamic adaptive auto-configuration;

– long life.

---

4 HQL: *Hibernate Query Language*.

TinyDB has been deployed in some major projects, such as environmental monitoring in *Great Duke Island* or in *James Reserve* in California.

Unlike conventional strategies that just aggregate and filter operations inside network, and limit the consumption of computing power and energy, *TinyDB* uses a control strategy to master where, when and how to capture and transmit data. "Acquisitional techniques " (ACQP) are involved in different parts of the system: (*i*) query optimization; (*ii*) dissemination of elements of queries on network; (*iii*) execution of queries on a sensor network.

The sensors are coupled to a battery and a very small and low-energy processing unit, on which the *TinyOS* operating system, which is itself scheduled to respect energy problems of embedded sensors, is installed. All this system is called a *Mote* or a node of network. *TinyOS* is specifically designed to handle hardware and communicate with other *Motes*. The programming is done using the *nesC* language, which is also specially created to address energy problems. *TinyOS* is very light, without kernel, or particular service of management or memory protection. It has a library to manage communication packets, read values from sensors, synchronize clocks and effectively disable hardware. *Motes* communicate by radio with each other via an antenna and a transmitter.

The architecture of the sensor network is dynamically built. A strategy based on spanning tree is generally used to enable communications (a root node is elected and then sends a query to his son, who in turn transmit it to his son, and so on). A network node chooses his father (the node that precedes it in the spanning tree) closest to the root and uses an estimate of the connection quality.

With *TinyDB*, acquisitional features, such as the capacity of the system to dynamically reconfigure, the processing inside network, the multi-hop routing, the discovery of the topology of communication network and the knowledge of the routing tree can be simply programmed by its declarative language which is similar to SQL (SELECT, FROM, WHERE, GROUPBY, etc.). The main difference as compared to SQL is that *TinyDB* constructs a query stream and a response stream. A table of *sensors* is distributed over all nodes that generate and store data. For reasons of economy, the records are made if necessary and stored for a very short time. Materialization points on each *Mote* are used to store data and perform flow windowing, enabling functions such as sorting or aggregation.

All queries have the following form:

SELECT select–list
[**FROM** sensors]
WHERE where–clause
[**GROUP** BY gb–list

[**HAVING** having–list]]
[**TRIGGER ACTION** command–name[(param)]]
[EPOCH DURATION **integer**]

To save energy, *Motes* work in interruption mode, and not in scanning mode, to keep them in an idle state as long as possible. Events are reported only in locally and do not wake up all *Motes*, but the signalling can be propagated through the affiliations along the spanning tree. The lifetime is regularly estimated and *TinyDB* adapts regularly transmission rate and sampling rate to satisfy an initially determined battery lifetime. For this, *TinyDB* sends monitoring queries to control the network status and exploratory queries to find out status of a node.

Queries can also act and generate an event, or require a storage in EEPROM memory. To optimize each query, *TinyDB* uses operational research techniques on series-parallel graph to find the order of predicates of the query which will result in minimal energy expenditure. To optimize queries in a global way, they are converted into the event flow and rewritten to avoid sampling the same sensor, multiple times.

To reduce energy, *TinyDB* builds for each query a routing tree which will awaken the least number of *Motes* as possible, knowing that to reach a *Mote*, all of his ancestors must be woken up. The choice of a new parent is done according to various pre-defined policies. Finally, in query processing, a queue is available on each *Mote* and policies (FIFO, "averaging heading tuples", etc.) and are designed to optimize the use of these files.

Considering the Internet of Things, these techniques can be applied when one wants to measure the environment, such as the temperature in a cold storage. In this case, passive RFID tags must be replaced by active tags equipped with a sensor. Passive RFID tags do not have energy problems and do not provide information other than its code that identifies if and the data stored in its memory. In this case, the sensor provides the value of a quantity, which evolves with time and requires the use of a battery. However, one can find these issues of energy consumption in the absence of sensor, if we consider active tags to emit at longer distances than those typically used with passive tags.

9.5.2. *GSN*

Global Sensor Networks [GSN 09] (GSN) is a software project started in 2005 at EPFL (*Ecole Polytechnique Fédérale de Lausanne*), in *Laboratoire de Systèmes d'Information Répartis* (LSIR) by A. Salcher under the supervision of K. Aberer. The initial goal was to create a reusable software platform for processing data streams generated by wireless sensor networks. The project was a success and was later reoriented to a platform of generic stream processing. GSN is an *Open Source* project composed of 3 parts:

– data acquisition;

– data processing (filtering with an enriched SQL syntax, algorithm execution on results of queries);

– data restitution.

GSN can be configured to acquire data from different sources: devices running on *TinyOS*, serial devices, communications with UDP, webcam, memory, etc. A large number of sources allows sophisticated data processing scenarios. If sources are not supported, it is easy to develop a Wrapper to operate new equipments with GSN. GSN offers advanced data filtering features based on an enriched SQL syntax.

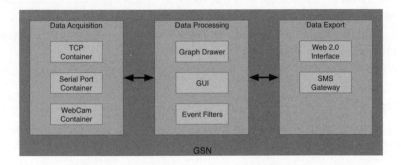

Figure 9.10. *GSN and three independent components*

GSN considers two types of data sources:

– data related to events. They are sent by the source and a GSN method is called when they arrive. Serial ports, network connections (TCP or UDP) correspond to this event logic;

– data related to the scanning. GSN periodically asks the source of new data. It uses RSS flow or an email account corresponding to this scanning logic.

The reception of data sent by different sources is possible with the use of *Wrappers*, which encapsulate data received from source in a GSN standard data model, called *StreamElement*. Concretely, a *StreamElement* is an object representing a row in an SQL table. Each *Wrapper* is a Java class that extends the parent class `AbstractWrapper`. A *Wrapper* initializes a tiers library, at the call of its constructor. This library provides a method that is called each time the library receives data from the monitored device. This method will extract relevant data and create one or more `StreamElement` with one or more columns. The received data is matched with an SQL data structure, with fields that have the same name and type. GSN will be able to filter data with an enriched SQL syntax.

Currently there are twelve available *Wrappers* whose list is given below, but there are possibilities to develop further other ones:

– *Remote Wrapper*, that allows a virtual sensor to use another virtual sensor as a data source;

– *TinyOS 1.x Wrapper*, which makes it possible to communicate with any device running OS *TinyOS* 1.x;

– *TinyOS 2.x Wrapper*, which makes it possible to communicate with any device running OS *TinyOS* 2.x;

– *Serial Wrapper*, which reads and sends data to the serial port (RS-232), real or virtual (wireless Bluetooth link);

– *UDP Wrapper*, which opens a UDP *Socket* on a configured port that reads the above data;

– *System Time Wrapper*, which generates events from a system clock every t μs, within the system time at which the event was generated;

– *HTTP Get Handler Wrapper*, which scans sensor readings from a remote *Web* server (like an IP camera);

– *USB Webcam Wrapper*, which scans periodically a USB camera to take an image;

– *Memory Monitor Wrapper*, which generates periodically usage statistics from memory.

GSN provides two complementary mechanisms for data processing:

– event counting and time cutting in windows, with the enriched SQL syntax;

– data manipulation due to the execution of predefined virtual sensors, which can be used, for example, to generate graphs, or a graphical interface to control devices, etc. We could also program their own virtual sensors. All programs are written in Java.

The data restitution is a domain which is still under construction, and it must be achieved using a *Web* interface, by mails, SMS, an integration into a Google map (especially when we want to make the location), etc.

The advantages of GSN are the following:

– economy of material resources of sensor networks;

– simplicity of communication with the best known *Motes* of the market which integrate technologies to sensor networks, such as the *TinyOS* operating system. It also provides a simple way to communicate with others by writing *Wrappers* in Java;

– ability to filter and process data from sensors, and an easy way to program and customize them;

– low memory consumption authorizing the installation in mobile systems like mobile phones. The data restitution can be done on a mobile phone or a Web interface.

GSN has the following disadvantage:

– low-level programming;

– early configuration;

– absence of system with very advanced configuration, reflexivity, behavioral or business analysis, more elaborate scenario, fault tolerance, failure tolerance, etc. But this may suffice in some uncritical cases and may even be required to comply with constraints of energy resources, moreover as another middleware can support these other aspects.

9.6. *ObjectWeb* projects and the Internet of Things

9.6.1. *Presentation of ObjectWeb [OW2 09a]*

It is an international non-profit consortium founded by BULL, INRIA and France Telecom in 2002, dedicated to the development of *Open Source* middlewares. It brings together enterprises and research organizations such as INRIA, Bull, France Telecom, Thales Group, NEC Soft, Red Hat or SuSE. This consortium has 70 companies as members and about 5,000 developers. The offer of *ObjectWeb* complies with the standards established by independent bodies such as JCP, OMG or OSGi. *ObjectWeb* federates more than 120 projects in various fields such as electronic business, grid computing, manufacturing messaging, *Web* services, etc. Among the most important projects noted were: *JOnAS* (Java application server certified J2EE) and JORAM (asynchronous message bus) which will be presented in this section.

9.6.2. *JORAM, component of ObjectWeb RFID*

JORAM [OW2 09b] is an asynchronous communication middleware that implements the JMS specification. This is an *Open Source* implementation (LGPL) which is fully based on Java. It meets the latest JMS 1.1 specifications and is part of the J2EE 1.4 specification. It is used in many operational environments, where it is employed as:

– autonomous Java messaging system between JMS applications developed for various environments (from J2EE to J2ME).

– messaging component integrated into a J2EE application server. As such, it is an essential block of the J2EE *JOnAS* server, which will be presented later.

JORAM consists of a server which manages JMS objects (*Queues*, *Topics*, *Connexions*, etc.) and a JORAM client part, where the JORAM client part is associated with a JMS client application. The architecture of JORAM is illustrated by Figure 9.11.

The communication between server and client relies on the TCP/IP protocol. An alternative is to use the HTTP/SOAP protocol for the JMS clients developed in a

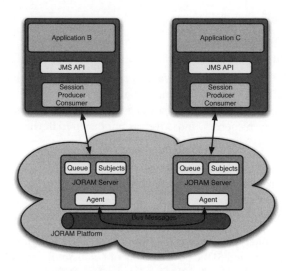

Figure 9.11. *JORAM platform*

J2ME environment. Communication between two servers can use different protocols depending on needs (TCP/IP, HTTP, SOAP, secure communication via SSL). Clients and servers may or may not be on different physical machines, and also may or may not run in different processes.

9.6.3. *Architecture of JORAM*

The essential feature of the JORAM platform is its distributed and configurable architecture. The basic architecture is of *Snowflake* type, i.e. consisting of a set of distributed JORAM servers, interconnected by a message bus. Each server handles a variable number of JMS clients. The repartition of servers and the distribution of clients on servers are the responsibility of the platform administrator, according to the needs of the application. This choice forms a first-level configuration. A second level is the ability to locate destinations (*Queues* and *Topics*) as required. A final level is QoS parameters associated with message bus (communication protocol, security, persistence, etc.). The choice of mechanisms is based on a balance between QoS level and cost. JORAM offers two models of communication, point-to-point and publish-subscribe.

9.6.4. *Advanced functions of JORAM*

JORAM proposes a set of additional features that are not part of the JMS 1.1 specification.

9.6.4.1. *Load balancing*

The architecture of JORAM implements load balancing mechanisms to increase the availability, due to the replication of communication objects, and optimize the flow of messages between servers.

The implementation of *Topics* is based on a "clusterized" *Topic* which is replicated on strongly coupled servers (cluster of machines) and to geographically distributed servers. Each time a replication node serves one of its subscribers, it also sends a message to other replication nodes. The sharing of messages between nodes reduces the traffic. Using this scheme, it is possible to be fault tolerant because if a server fails, it does not affect the rest of the application.

The implementation of *Queues* is based on the existence of several copies of the same *Queue* object, located on independent servers. Each copy is available to customers connected to the server. If the load of a server exceeds a threshold, the messages are redirected to another copy of the same *Queue* managed by another server. This concept also improves the level of availability and performance, without impacting application programming.

9.6.4.2. *Reliability and high availability*

Mechanisms based on acknowledgement between a JMS client and its representative in the server (object *Proxy*) render communications reliable even though the exchanges are asynchronous. This solution is complemented by Store & Forward mechanisms conducted by *Proxy*, for exchanges between the *Proxy* and destination, and by a message bus for exchanges between servers.

For this, a version JORAM HA (*High Availability*) was designed, according to an active replication approach in master-slave mode. *Queues* and *Topics* are replicated on JORAM servers running on a cluster of machines. The master server executes queries of clients and propagates operations to the slave server that replicates the processing locally. The current version of JORAM HA uses JGroups [JGR 09] mechanisms during communication.

9.6.4.3. *Extended connectivity*

Connectivity can be enhanced by:

– a JMS gateway for the interoperability with other JMS platforms. The link is established through a destination object JORAM which represents the final destination;

– the use of the SOAP protocol, which provides a standard way to access remote services by exchanging XML messages on HTTP connections. This is useful to respond to security constraints imposed by the management of firewalls, or to take into account clients running in a J2ME environment for which the complete API JMS cannot be provided.

JORAM is compliant with the specifications of *J2EE Connector Architecture* (JCA), which describes how to integrate external resources into a server of J2EE applications. Thus, we can manage the lifecycle of resources, connections with EJB components, and perform transaction management according to the XA interface. This makes it possible to integrate JORAM to any server of J2EE application which implements this specification. It is in particular the classical integration path with the JOnAS server.

9.6.4.4. *Security*

JORAM uses on-demand SSL connections to authenticate participants and encrypt messages. Nevertheless, firewalls should be well configured for the use of ports. One solution is to use in the JORAM platform protocols commonly accepted by firewalls (HTTP and SOAP).

9.6.5. *Ongoing works on JORAM*

Several notable works are in progress to evolve JORAM toward issues which are specific to systems based on RFID tags such as:

– extend the application scope into the field of embedded systems. This objective requires a light version of the JORAM server to meet the resource constraints of many classes of devices such as smart card, RFID reader, industrial controller, etc., particularly for RAM memory and persistent memory like flash memory. One approach is to deport a *Store and Forward* function in the client library. This new structure would implement the P2P links between JORAM clients without using a tiers server;

– improve performance and administration functions. The idea is the use of a database management system to manage the persistence of messages, or in longer-term the use of self-adaptive mechanisms to enable a JORAM platform to gather information about its behavior to spontaneously adapt to failures, performance declines, or configuration changes.

9.6.6. *JINI technology and the Internet of Things*

9.6.6.1. *Generalities*

The aim of this technology is to make applications independent of operating systems. Considering peripheral devices and software as independent objects that can communicate, JINI can gather them into a federation of objects which are automatically installed and function as soon as they are connected.

JINI is a network architecture for building distributed systems in the form of modular co-operating services. Originally developed by *Sun Microsystems*, the responsibility of JINI was transferred to *Apache* under the project name *River*. JINI [JIN 09] provides solutions to address problems of system evolution, resilience, security and dynamic assembly of service components. Code mobility is a fundamental concept of the platform and provides an independence regarding communication protocols.

The technology, introduced by SUN in 1998, simplifies the interfacing and the connection between portable computers (PC, PDA, mobile phone, TV, game console, digital camera, printers, fax, alarms, GPS, home automation, etc.). In addition, this technology provides a framework in which components can make their existence and their service known. Any software can find a service that implements a set of data features and establish a user agreement.

9.6.6.2. *Architecture of Jini*

A service can be implemented in hardware (a printer offers a printing service), as well as in software (text processing can be seen as a service). In both cases, the technology is based on dynamic loading of code. This data transmission is done through a stack of protocol layers. Connection management is performed by a proprietary protocol JRMP[5] based on TCP/IP.

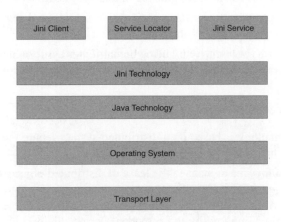

Figure 9.12. *Architecture of Jini*

The JINI technology is designed to be viewed as a network extension of infrastructure, programming model and services of the Java technology.

9.6.6.3. *Infrastructure of Jini*

There are three basic components in the infrastructure of JINI:

– Java RMI classes that allow us to manipulate remote objects and ensure the security of the JINI distributed environment;

---

5 JRMP: *Java Remote Method Protocol.*

	Infrastructure	Programming model	Services
Java	Java VM RMI Java Security	Java APIs Java Beans ...	JNDI Enterprise Beans JTS
Java + Jini	Discovery/Join Distributed Security Lookup	Leasing Transaction Events	Printing Transaction Manager JavaSpaces Services

Table 9.1. *Jini*

– Discovery Process and Join Process which respectively make it possible to discover a registering service and then register a service;

– Lookup Service which is the collection of existing services.

9.6.6.4. *Operation of Jini*

Two different roles can be distinguished in a JINI system:

– the service provider looks for a registration service and registers the objects making services to export and its attributes;

– the client looks for a service by using the attributes associated with it. A copy of registered objects is transmitted, so that it can communicate with considered service.

9.6.6.5. *Jini programming model*

The JINI technology is based on a distributed system reputed to be reliable. The model proposes three API:

– *leasing*: it allows us to manage the lease of distributed objects by extending the notion of *Garbage Collector* into distributed environments. The duration of the lease, i.e. the period during which the provider guarantees access to the requested resource can be negotiated or imposed by the provider. During this period, the lease can be cancelled by the applicant; resources are then released;

– *transaction*: operations are grouped, so as to extend the atomicity principle: transactions succeed or fail together. To an outside observer, they are a single operation. A transaction is created and monitored by a transaction manager which implements the *TransactionManager* interface. The validation of a transaction implies the vote of each participant. Three votes are possible: (*i*) *Prepared* if the participant estimates that the transaction can be validated, (*ii*) *Not Changed* if operations were only in reading, (*iii*) *Aborted* if the participant estimates that the transaction should not be validated;

– *events*: is an extension of the event model used by the *JavaBeans* component model extended to a distributed environment. JINI provides an event communication model between different services of the system.

The implementation of these APIs is done through the following steps:

– definition of interface for the remote class, which inherits from the *Remote* interface. Global public methods of the object are declared. These methods can emit the *RemoteException* exception;

– definition of the service class that implements the interface defined in the previous step and the *ServiceIDListener* interface, necessary for the recall of *JoinManager*, when a service identifier is assigned. This interface is used for services that have not had an identifier yet. Finally, the class must inherit from *UnicastRemoteObject* class or *Activatable* class. The objects manipulated or exchanged by the RMIserver must be *Serializable* to be transferred over the network;

– definition of a *Service Provider* of *ServerUniCast* type if the registration service (*LookupService*) can be localized on network, or of *ServerMultiCast* type if a *LookupService* must be searched in advance. It is the *Service Provider* who will instantiate the JINI service if it is necessary;

– definition of the client program that is similar to the server created in the previous step uses a *ClientUniCast* object if the registration service (*LookupService*) can be localized on network, or *ClientMultiCast* if the *lookupService* is unknown;

– generation of *Stub* using the `rmic` command;

– launching the RMIregister (`rmiregistry`) followed by the registration service of JINI (*lookupService*) services by using *start* archive provided with JINI;

– launching the *Service Provider*;

– finally, launching the client.

9.6.7. *JONAS, component of ObjectWeb RFID*

Developed within the *ObjectWeb* consortium, JONAS [OW2 09a] is a server of J2EE *Open Source* applications. JONAS, in particular provides support for EJB's and *Web* services, "clusterization" functions, integration of JMS and JASMINE administration tools. JONAS is usable on many operating systems, *Web* servers and *Open Source* or commercial databases. It received the J2EE certification in early 2005.

Within the ObjectWeb consortium, JONAS is developed among others, by the company Bull that manages the core of the application server and its evolution. Development teams are in Grenoble (France) and Phoenix (USA). Bull also contributes to the development of external components used in JONAS, such as Tomcat and Axis, and proposes complementary softwares to JONAS:

– *EasyBeans*: a J2EE container used on JONAS 4 and JONAS 5, on Tomcat and in standalone;

– *Bonita*: a workflow engine running on JONAS;

– *Orchestra*: a service orchestration engine;

– *ExoPlatform*: a portal which is deployed on JONAS;

– *WTP*: a *Plugin* for the *Eclipse* development environment.

In addition, Bull proposes an offer of professional support for JONAS and provision of services (training, development, integration and accommodation) around JONAS and J2EE.

The Distributed Systems Laboratory (DSL) of the Polytechnic University of Madrid participates in the development of component *Cluster* of JONAS which ensures the high availability and fault tolerance of J2EE applications. Contrary to other approaches of *Clustering* J2EE, JONAS ensures transactional consistency, even in case of errors. DSL works on new replication protocols and begins working on autonomous communication protocols for JONAS.

The University of Fortaleza has created a research lab focused on J2EE architectures and in particular EJBs, *Web* services and *Clustering*. Activities started in September 2006 by the implementation of a group communication (GC) layer for the architecture of *Cluster* JONAS 5.

OrientWare is a non-profit consortium founded by *BeiHang* University (BHU), Institute of Software, Chinese Academy of Sciences (ISCAS), National University of Defense Technology (NUDT) and Peking University (PKU) to integrate more than 800 Chinese projects in the field of middleware and disseminate scientific research results by using the *Open Source* mode as the primary distribution channel, in order to increase collaboration between academia and the world of industry. Some of the members of the Peking University work on the JONAS project, and it was recently decided to merge the application server of ObjectWeb, JONAS with the one at the Peking University (PKUAS). Since 2006, PKU contributes to JONAS on aspects of *Clustering*.

On its side, France Telecom R&D contributes to the EJB container of JONAS for the management of persistence and is the origin of *Java Object to Relational Mapping* (JORM) used for the support CMP2 in JONAS and in SPEEDO (JDO implementation). France Telecom also provides an implementation of EJB3 the persistence (JPA).

The *Laboratoire d'Informatique Fondamentale de Lille* (LIFL) contributes to JONAS on a tool of deployment (FDF) and for monitoring (*Thread Management Framework*). This work is partially funded by the European project ITEA S4ALL.

JONAS adopts an OSGi architecture, offering modularity and flexibility. It is implemented as a set of OSGi *Bundles*, with services in the form of OSGi services. The server architecture enables us to add new services or replace existing services by alternatives. Services can be started, stopped and reconfigured at runtime.

9.6.8. *ASPIRE initiative of OW2*

The project ASPIRE[6] [ASP 09a] RFID is defined as a project "to develop and promote a middleware (LGPL v2.1) that is open-source, lightweight, standards-compliant domain, capable of scaling to large numbers of items, considers the respect of privacy as well as a series of tools to facilitate development, deployment, application management based on RFID tags and sensors". ASPIRE implements several specifications produced by different consortia in the field such as EPCglobal, *NFC Forum*, JCPs and the OSGi alliance.

It is a project of the *ObjectWeb* which participates in a European initiative in the development of RFID tags [ASP 09b]. In practice, it is a complete environment that is provided, and is capable of handling different types of readers and RFID tags.

9.6.8.1. *Architecture*

The ASPIRE architecture (Figure 9.13) offers a first abstraction layer above hardware: the HAL[7] layer. Among supported equipments, we initially find equipment which are compliant to the EPCglobal standards (using RP[8] and LLRP[9] protocols). In addition, we also equipment compliant to NFC standards and especially mobile phones equipped with a NFC system. The communication between user applications and readers themselves is standardized and uniform. Readers non compliant to the EPCglobal protocols are managed through a *Proxy* making them compatible with these protocols.

The upper layers of the architecture reuse the EPCglobal specifications. An intermediate layer corresponding to the ALE (section 6.3) is present and is complemented by a component called *Business Event Generator* (BEG). As specified in the EPCglobal standard, to minimize the data exchanged by many readers and tags, a mechanism for filtering and aggregation is implemented (*Filtering and Collection Layer*) which attempts to reduce the volume of traffic as soon as possible, i.e. as close as possible to hardware. On its side, BEG tries to intervene just after the filtering and to send and structure information with the greatest relevance for the business concern of the application. The idea of this component is to clearly separate business-oriented issues, aspects related to hardware and in particular the amount of exchanged information. BEG formats information to generate EPC-IS events which will be sent to storage systems of the EPC-IS information system.

---

6 *Advanced Sensors and lightweight Programmable middleware for Innovative Rfid Enterprise applications.*
7 *Hardware Abstraction Layer.*
8 *Reader Protocol.*
9 *Lower Level Reader Protocol.*

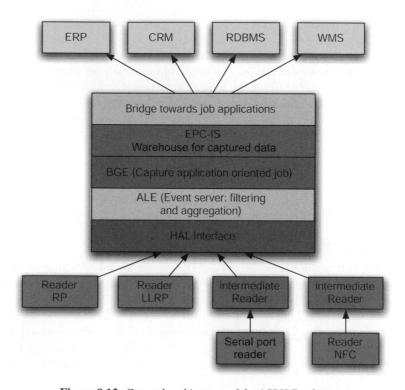

Figure 9.13. *General architecture of the ASPIRE solution*

Storage systems of EPC-IS data will then store information from BEGs through capture interfaces compliant to the EPCglobal standard and will be able to restore the data through the query interface besides via a number of gateways to other definition mechanisms for enterprise information systems such as ERPs, CRMs, etc.

9.6.8.2. *Deployment*

The ASPIRE architecture is deployed on a physical infrastructure of type EPS[10] [HA 08].

The *Edge* part is hosted by an OSGi platform which is responsible for management of RFID readers. The role of this gateway is the collection of raw data from readers and tags. These gateways embed essentially control modules of readers and the HAL part of the ASPIRE architecture.

---

10 *Edge-Premise-Server.*

The *Premise* part is also hosted by an OSGi platform and is also primarily used to administrate readers and secure *Edge* platforms. This part of the infrastructure can also support functional aspects by integrating mechanisms for managing relationships between multiple *Edge* gateways and functions of aggregation and filtering. This allows the deportation as early as possible (closest to the source) the flow of information, and basic mechanisms of the ALE component of ASPIRE architecture.

Finally, the *Servers* part is hosted by an application platform of type JBOSS or JONAS. It houses ALE and BGE components, as well as EPC-IS depots and various software gateways to enterprise applications. This part of the infrastructure can be deployed centrally or in distributed manner.

ASPIRE also provides a mechanism to fully administer the deployed solution by using the JMX[11] technology. Thus, all components of the solution expose a standard JMX interface enabling the development of complex monitoring applications to exploit both hardware and software components and associated servers.

9.6.8.3. *Development*

One of the key points of the ASPIRE proposal is the provision of an integrated development environment (IDE) facilitating the control of different creation phases of an RFID solution. The objective of this IDE is to enable a precise specification of requirements to automatically generate a solution based on ASPIRE technology choices. IDE must thus facilitate:

– configuration of physical readers;

– configuration of associated HAL layer;

– server management of the ALE layer and in particular the component of filtering and aggregation;

– editing commands compliant to the ALE standard for the component of Filtering and Aggregation;

– specification of businesses-oriented information flow;

– management of business-oriented tiers application connectors;

– editing meta-data specific to enterprise.

Therefore, different modules allow us to specialize behavior of the main components of the deployed platform and in particular at the BEG that captures the business expertise of the application.

The ASPIRE project highlights European ambitions to overcome the stranglehold of the solutions from U.S. leading manufacturers. Although based on the EPCglobal

---

11 *Java Management eXtension.*

standard, this solution aims at providing European companies with an access to a complete solution, *Open Source* and free of copyright.

9.7. Conclusion

An initial observation leads us to think that some of the solutions are rather of the "announcement effect" rather than a real development of the RFID technology. Most major participants have adapted their business solutions at lower cost to position themselves in a promising market. However, the late arrival of this technology does not really encourage them to invest in large specific developments. The vast majority of these solutions follow the same general architecture, based on the EPCglobal standards.

It is also interesting to note that major vendors have preferred to build solutions based on individual requirements by allowing customers to test them on pilot sites. This approach shows us that RFID technology still needs conviction (Chapter 4).

In parallel, *Open Source* and academic solutions have emerged, depicting a growing interest in these technologies. To have an important position, RFID technologies must answer two very important points:

– ensure the interoperability of solutions to enable various systems based on different technologies to work together and distributed applications to play their role. In fact, like the classic Internet, the Internet of Things will only work if it guarantees the possibility to any network element to interact with any other element, taking into account the imperatives of different security policies;

– provide an integration at low cost is a major challenge that requires a technology evolution at the level of tag production but also in terms of their inclusion in information systems of enterprises. In addition, it cannot be done to the detriment of security imperatives.

Current middleware solutions bet on the integration through the conformity of the EPCglobal standard. However, this conformity is sometimes an obstacle to the expansion of the use of RFID tags, because the EPCglobal model and the need to pass by its network, are reasons for suspicion for companies regarding these solutions. It is still too early to comment on the long-term success of these technologies; but it seems that it can be done without taking into account at all levels, and especially the middleware security aspects (confidentiality, integrity, privacy, etc.).

9.8. Bibliography

[ASP 09a] OW2 Consortium, ASPIRE RFID wiki, 2009, http://wiki.aspire.ow2.org.

[ASP 09b] European Union Project,   ASPIRE, University of Ariborg, Denmark, 2009, http://www.fp7-aspire.eu.

[GSN 09]   Sourceforge, GSN, 2009, http://sourceforge.net/apps/trac/gsn/.

[HA 08]   HA T.-T., Déploiement automatisé d'architectures Edge- Premise-Server (EPS) dans le contexte des intergiciels RFID, Master's thesis, University Joseph Fourier, June 2008.

[HIB 09]   Hibernate, 2009, http://www.hibernate.org.

[IBM 09]   "IBM Sensors and Actuators", 2009, http://www-01.ibm.com/software/solutions/sensors/.

[JBO 09]   JBoss Community, 2009, http://www.jboss.org/.

[JGR 09]   JGroups - A Toolkit for Reliable Multicast Communication, 2009,   http://www.jgroups.org/.

[JIN 09]   "Jini Specifications and API", Sun Microsystems - Product, 2009, http://java.sun.com/products/jini/.

[OW2 09a]   OW2 Consortium, JONAS, 2009, http://wiki.jonas.ow2.org.

[OW2 09b]   OW2 Consortium, JORAM, 2009, http://joram.ow2.org.

[OW2 09c]   OW2 Consortium, ObjectWeb, 2009, http://www.ow2.org.

[RFI 09]   Microsoft BizTrik, RFID help, 2009,   http://msdn.microsoft.com/en-us/library/dd352559(BTS.10).aspx/.

[SIK 09]   Sikander J., "Microsoft RFID Technology Overview", MSDN, 2009, http://msdn.microsoft.com/en-us/library/aa479362.aspx/.

[SIN 09]   Singularity, 2009, http://http://singularity.firstopen.org/.

[SUN 05]   SUN MICROSYSTEMS, "The Sun Java^{TM} System RFID Software Architecture", Technical White Paper, March 2005,   http://www.sun.com/solutions/documents/white-papers/re_EPCNetArch_wp_dd.pdf?facet=-1.

[SUN 09]   Oracle, "http://www.sun.com/", 2009.

[TIN 09]   "TinYdB", 2009, http://telegraph.cs.berkeley.edu/tinydb/.

[WEF 06]   OASIS Standard,   WSDM V1.1, August 2006,   http://www.oasis-open.org/home/index.php.

List of Authors

Stefan BARBU
NXP Semiconductors
Zurich
Switerland

Mathieu BOUET
LIP6
Paris
France

Julien BRINGER
Morpho
Osny
France

Hervé CHABANNE
Morpho & Télécom ParisTech
Issy-Les-Moulineaux & Paris
France

Daniel DE OLIVEIRA CUNHA
LIP6
Paris
France

David DURAND
IUT
Amiens
France

Simon ELRHARBI
Telecom ParisTech
Paris
France

Vincent GUYOT
ESIEA LIP6
Paris
France

Thomas ICART
Paris
France

Yann IAGOLNITZER
Education Nationale
CNED
Boussac-Bourg
France

Patrice KRZANIK
Conservatoire National des Arts et Métiers
Paris
France

Thanh-Ha LE
Morpho
Osny
France

François LECOCQ
Morpho
Osny
France

Christophe LOGE
University of Picardie
Amiens
France

Dorice NYAMY
Telecom ParisTech
Paris
France

Cyrille PÉPIN
Morpho
Osny
France

Guy PUJOLLE
LIP6
Paris
France

Jean-Ferdinand SUSINI
Conservatoire National des Arts et Métiers
Paris
France

Pascal URIEN
INFRES
Télécom ParisTech
Paris
France

Index